FIFTY YEARS OF THE OUTER SPACE TREATY

Tracing the Journey

FIFTY YEARS OF THE OUTER SPACE TREATY

Tracing the Journey

Editor

Ajey Lele

idsa

INSTITUTE FOR DEFENCE STUDIES & ANALYSES
NEW DELHI

PENTAGON PRESS

Fifty Years of the Outer Space Treaty: Tracing the Journey

Editor: Ajey Lele

First Published in 2017

ISBN 978-81-8274-948-1

Published by
PENTAGON PRESS
206, Peacock Lane, Shahpur Jat
New Delhi-110049
Phones: 011-64706243, 26491568
Telefax: 011-26490600
email: rajan@pentagonpress.in
website: www.pentagonpress.in

In association with
Institute for Defence Studies and Analyses
No. 1, Development Enclave,
New Delhi-110010
Phone: +91-11-26717983
Website: www.idsa.in

Printed at Aegean Offset Printers, Greater Noida, U.P.

Contents

Foreword

Since the day the first man-made satellite, the Sputnik, spun into earth's orbit nearly six decades ago, the international community embraced the idea of the use of outer space exclusively for peaceful and scientific pursuits. This soon became the established global norm the very first time the United Nations General Assembly considered the 'Questions of the Peaceful Use of Outer Space' in 1958 and encapsulated it in its resolution 1348(XIII). Conscious that space exploration had opened new possibilities for the improvement of life of humankind, the General Assembly also created the Ad-hoc Committee on the Peaceful Uses of Outer Space (COPUOS) to harness outer space activities for cooperative mutual gain. It is remarkable that this happened notwithstanding the heightened Cold War competition between the United States and former Soviet Union.

COPUOS thereafter developed five treaties, including the Treaty on the Principles Governing the Activities of States in the Exploration and Use of Outer Space, including the Moon and other Celestial Bodies, better known as the Outer Space Treaty, which constitutes the cornerstone of the international legal framework for the peaceful application of outer space. Its four core principles are that the exploration and use of outer space shall be carried out "for the benefit and interest of all countries", that outer space will be "the province of all mankind", that outer space shall be "free for exploration and use by all countries", and that the States Parties of the Treaty undertake not to place in orbit "any object carrying nuclear weapons or any other weapons of mass destruction" in the orbit of the earth, moon or any other celestial body. It further enjoins the States Parties to use these bodies exclusively for peaceful purposes, not to subject these to weapons testing or military manoeuvring, and prohibits the establishment of military bases, installations and fortifications thereon.

Indeed, since Sputnik's launch, the peaceful applications of space technology, especially for communications, education, navigation, and meteorology, have contributed significantly to global growth, besides assisting emerging economies such as India to solve problems of sustainable national development. As with technology in other domains, space technology has dual use features, which lends it to defence applications. It ensures strategic stability by enabling effective intelligence, surveillance, and reconnaissance capabilities. Space situational awareness for early information, high-resolution sensors and imagery to map ground-based offensive systems, and precision guidance for weapons delivery are some of the space based defence applications.

Indian satellites in outer space establish global connectivity, and help in eradicating illiteracy, providing health security, improving navigation and meteorological services, optimising management of natural resources and the environment, and coping with extreme weather events and natural disasters. An innovative application has been the setting up, across India, of Village Resource Centres as a single-window delivery mechanism for a variety of space-enabled services, including tele-education, tele-medicine and interactive advisories on land and water management.

There has, thus, been a dramatic acceleration, in recent years, in the peaceful uses of outer space and in international cooperation for this purpose. So also has there been an increased potential, in particular for the developing countries, to leapfrog and become full participants in the technology-based global economy of the twenty-first century. Given the increasing dependence globally on the use of outer space for development purposes, and the all-pervasive application of space technology for almost every aspect of modern life, it is imperative to guarantee the security of assets based in outer space, for any threat to them would disrupt space-enabled services to all countries, irrespective of the level of their development.

Although the Outer Space Treaty has largely ensured responsible conduct of space activities, the prohibitions contained in the Treaty do not apply to the transit of weapons of mass destruction through outer space if launched by ballistic missiles, or the use of conventional weapons towards, through, and from space. By not unambiguously enumerating the permitted or prohibited activities, it provides an imperfect prescription of the legal obligations and limitations imposed on the States Parties in respect of their offensive activities in outer space, including the development, testing, and possible deployment of kinetic energy Anti-satellite Weapons in outer space designed to destroy space-based assets.

India has been unequivocal in its opposition to preventing the weaponisation of outer space. P.V. Narasimha Rao, as Foreign Minister, had suggested in the early 1980s that an arms buildup in outer space would "mean a permanent goodbye to disarmament and peace and plunge mankind into a perpetual nightmare." Thereafter, India's unequivocal stand against the weaponisation of outer space continued to be enunciated in national and multilateral forums, including by former President Dr. A.P.J. Abdul Kalam, a well-known space scientist. Addressing a 'Space Summit' of the Indian Science Congress nearly twenty years later, he had cautioned the world's space community "to avoid terrestrial geo-political conflict to be drawn into outer space, thus threatening the space assets belonging to all mankind."

To this day, India has remained strongly supportive of the quest to upgrade the present international legal framework for regulating space activities, set at the relative infancy of the development of space technology, and to strengthen the existing space law for the peaceful use and exploration of outer space. The respect for the safety and security of space assets and capabilities of all countries is a prerequisite for ensuring the continued flow of space-enabled services to all countries, including to developingcountries.

The Final Document of the First Special Session of the UN General Assembly devoted to Disarmament (SSOD- I) had stipulated that, in accordance with the spirit of the Outer Space Treaty, further measures should be taken and international negotiations held in order to prevent an arms race in outer space (PAROS). As a consequence, the issue has remained on the agenda of the Conference on Disarmament since 1982, and an Ad-hoc Committee on PAROS functioned for a decade since 1985, mandates to examine, as a first step at that stage, "through substantive and general consideration", the issues relevant to the prevention of an arms race in outer space.

While this exercise did not produce substantive results, the issue remains relevant today, if not more than in the 1980s. Since then, both Russia and China have tabled papers and proposals in the Conference on Disarmament that enable a better understanding of the different dimensions of outer space security, and thus provides a good basis for commencing multilateral negotiations on PAROS, which have not so far materialised.

We are exactly fifty years from the date the Outer Space Treaty was opened for signatures in New York. This book attempts to contextualise the Treaty and examine its relevance in the 21st Century, while tracing its journey over the preceding five decades. The experts who have contributed to the book analyse key aspects of the legal arrangements enabled by it, as also their strength and limitations. Besides, the book presents views from a few important space-faring States concerning their role on taking forward this Treaty mechanism. I thank all the contributors for their valuable insights.

Outer space must be seen as part of the global commons, just as the oceans, since the destruction of space-based assets by weaponisation threatens both development and security. This is as good a time as any to buttressing the outer space legal regime by taking forward the idea of preventing an arms race in outer space.

January 27, 2017

Jayant Prasad
Director General
Institute for Defence Studies and Analyses

About the Contributors

Prof (Dr) Ram S. Jakhu, is a Director of the Center for Research of Air and Space Law, McGill University, Montreal, Canada.

Prof (Dr) Joan Johnson-Freese, is a member of the faculty of the Naval War College, Newport, USA.

Dr G.S. Sachdeva, is an Adjunct Professor at NALSAR, Hyderabad, India and specializes in International Air and Space Law.

Dr Ranjana Kaul, is a partner at Dua Associates, Advocates & Solicitors, New Delhi, India and specializes in Aviation Law and Regulations and Space enabled business regulations.

Dr Ji Yeon-jung, is a lecturer in Graduate School of International and Area Studies, Hankuk University of Foreign Studies (HUFS), Seoul, South Korea.

Prof (Dr) Frans G. Von der Dunk, is a Professor of Space Law at the University of Nebraska-Lincoln, Lincoln, USA.

Colonel Philip A. Meek (Retd, United States Air Force), Attorney, Austin, Texas.

Mr Aleksandr Klapovskiy is a retired diplomat. Formerly he was an employee at the Ministry of Defence in the USSR.

Mr Vladimir Yermakov is the deputy Director General of the Department for Non Proliferation and Arms Control, Ministry for Foreign Affairs of the Russian Federation.

Prof Li Juqian is a Professor at the International Law School at the China University of Political Science and Law, Beijing, China.

Mr Kumar Abhijeet, is an Assistant Professor, National Law School of India University, Bangalore, India.

Mr Munish Sharma, Associated Fellow, IDSA, New Delhi, India.

Dr Malay Adhikari, is an Assistant Professor at the School of Law, Faculty of Science and Humanities, SRM University, Kattankulathur.

Dr Eligar Sadeh, is a President and C.E.O. of Astroconsulting International LLC and also serves as the chief editor of the journal Astropolitics.

Dr Rajeswari Pillai Rajagopalan, is a Senior Fellow and Head of the Nuclear and Space Policy Initiative at the Observer Research Foundation New Delhi, India.

Introduction

Humans have been fascinated with outer space for centuries. There could be some mythological references offered in regards to the human activity in space, however, recent history of the human interest in the space domain could be said to have begun with the launch of the Sputnik satellite by the erstwhile USSR in 1957. Subsequent developments in the fields of rocket and space sciences have increased human interest in space multi-fold. Investments in space have provided various social and economic benefits. The overall process of space exploration has required scientists to make various technological innovations, which have also found utility in various other fields of life.

However, there are restrictions on the development of certain technologies, and in addition, rocket technology is an inherently difficult technology. Hence, today only a limited number of states – commonly called the spacefaring states – have demonstrated and dynamically used their own technologies for putting satellites in orbit (either in the low earth orbit or higher orbits). There are 11 such states, which include: Russia (the erstwhile USSR), Ukraine (being a successor state), the US, Japan, China, India, Israel, Iran, North Korea, South Korea and the European Union (EU). The other states – referred to as non-spacefaring states – are also keen to establish independent space programmes. Their main interest is to own a satellite in space, and for that purpose they depend on spacefaring states to provide launch facilities. It must be noted that private players are also entering the space arena with their own space programmes, and with few agencies having independent satellite launch facilities. Only three spacefaring states, namely the US, Russia and China have proven human spaceflight capabilities.

With satellite technology becoming more accessible and affordable, many states are investing in their first satellite system. Euroconsult's[1] 2016 report, *Trends and Prospects for Emerging Space Programs*, identified 24 countries as emerging space programmes (ESPs) in 2015. By 2025, it is estimated that the number of ESPs will increase to 47 countries around the world. This would include 23 newcomers who would have committed their first investment in space between 2016 and 2025.[2] As per some unsubstantiated reports, there could be even 70 states having interest (and keen to develop independent space programmes) in the space arena.

Since the beginning of the space era (1957), space has been viewed as a domain of civilian and military doings. However, conduct of activities in space has always been a major technological challenge. Various actors using satellites, for communication, navigation and remote sensing purposes, have found that these applications have utility for militaries too. Hence, there have been efforts to ensure that the space domain remains peaceful (in spite of being used for military purposes) and all nation-states get unrestricted access to space. Majority of states expect equal and non-discriminatory access to outer space and are of the opinion that technological supremacy of particular states should not lead to their domination. Overall, the medium of space should remain a cooperative endeavour, and all states should be able to derive various benefits.

The need for a globally accepted mechanism for the conduct of activities in space has been felt since the beginning. There have been efforts to establish a mechanism for the conduct of operations in space which is acceptable to all the stakeholders. The expanse of space being so vast, it has also been felt that certain categories require separate mechanisms. During the initial phase of the Cold War era, the US in the context of space activities argued that "peaceful" meant non-aggressive, while for the Soviets "peaceful" meant non-military.[3] But with the development of newer technologies, many states are looking for amity in space. States understand that weaponising space is of no use for anybody. Space actually needs to emerge as a common heritage for mankind and not a medium for warfare. Today, there is a general agreement, at least in principle, that various activities in space need be to undertaken under a rules-based approach.

During the Cold War era, multilateral disarmament and arms control negotiations were to some extent successful in avoiding the conflict and reducing the unnecessary arms build-up. During the 1960s, missile technology was central to the development of satellite launching systems. Also, satellites were viewed as: one, a demonstrator of technological superiority, and two, an instrument for intelligence gathering mainly in respect of nuclear aspects. Hence, major powers even then realised the need for the development of some legal guidelines/polices relating to outer space.

Interestingly, within 10 years after the launch of the first satellite, a treaty mechanism was agreed upon. Overall, in a short span of 12 years (1967-1979), the existing five treaties governing principles and details of universal co-operation in outer space were adopted. They were: the Treaty on Principles Governing the Activities of States and the Exploration and Use of Outer Space, Including the Moon and Other Celestial Bodies (Outer Space Treaty) of 1967; the Rescue and Return Agreement of 1968; the Liability for Damages Convention of 1972; the Registration Convention of 1975; and the Agreement Governing the Activities of States on the Moon and Other Celestial Bodies (Moon Treaty) of 1979. In sum, all these five mechanisms together provide for a remarkable network of rules governing a broad spectrum of issues peculiar to space.[4]

Amongst the above treaty mechanisms, the opening mechanism for space domain, the Outer Space Treaty (OST) is considered to be a holistic treaty (in a relative sense) on space issues addressing many important issues. It is important to note that the two nuclear superpowers, the US and the erstwhile USSR, entered into an agreement and signed this treaty at a time when the Cold War was at its peak. Possibly, both these powers recognised the potential damage they could have caused by extending their war into space, and hence bargained to stabilise the climate of conflict.

The OST was adopted by the United Nations General Assembly on December 19, 1966 and signed at Moscow, London and Washington on January 27, 1967. The treaty is largely based on the Declaration of Legal Principles Governing the Activities of States in the Exploration and Use of Outer Space, which had been adopted by the General Assembly in its resolution 1962 (XVIII) in 1963, with few additional provisions.[5] Further:

> This is the first multilateral convention to enumerate widely accepted guidelines designed to temper the intensity of potential disputes certain to arise in future allocation of both the spatial and material resources of outer space. The treaty attempts to conserve, as well as extend, mutually held values through the promotion of three basic objectives. Each article of the Treaty is subsumed under one of these three objectives:
>
> - To guarantee that outer space, including the moon and other celestial bodies, remains the heritage of all mankind.
> - Provides for cooperation among, and liability of, the parties exploring space.
> - To prevent the arms race from spreading to outer space. This objective exemplifies the widely shared values of security and well-being.[6]

The treaty mechanism entered into force as an international legal instrument on October 10, 1967. Over the years, the OST has gained international acceptance, and as of January 2017, 104 countries are parties to this treaty, while another 24 have signed the treaty but have not completed the ratification process. Moreover, many OST provisions are so well-established and well-observed that they are said to reflect "customary" international law and are thus binding even on those states which have not legally signed the treaty.[7] The year 2017 marks the completion of 50 years of the OST.

The OST has 17 articles and is based on "principles", and is therefore clearly intended to be neither exhaustive nor comprehensive. The process of treaty negotiations ensured that the parties agreed to the governance framework for those activities they could mutually agree on.[8] In the present context, since technology has leapfrogged during the last five decades, hence the world faces challenges which could not have been anticipated then.

It is important to appreciate the role played by the United Nations since the very beginning of the space era. In 1958, shortly after the launching of the first artificial satellite, the General Assembly established an ad hoc Committee on the Peaceful Uses of Outer Space (COPUOS). In 1959, the General Assembly established COPUOS as a permanent body, and today it has 84 members. In addition to state parties, a number of international organisations, including both intergovernmental and non-governmental organisations, have been granted observer status with COPUOS and its subcommittees. Actually, the OST itself was negotiated within COPUOS. Some of the United Nations documents/declarations help in highlighting the role played by various agencies towards assessing their own commitments to the treaty.

This edited volume traces the journey of the OST over the last five decades. It presents the views of experts on the role played by the treaty so far and its relevance for the present and future. Various chapters discuss a range of issues – from the evolution of the treaty, its strengths and limitations to its relevance in the present context and the future. Few chapters also discuss the efforts made by individual states during these decades towards strengthening this mechanism.

During the last 50 years, depending on the situations then prevailing, some useful suggestions emerged that have made the treaty mechanism more operative. At the same time, efforts are also being made to formulate a new structure for addressing various challenges in space. A large number of areas of terrestrial law are relevant to outer space, too. Primarily, laws of sea are particularly significant in shaping various policy choices for the conduct of activities in outer space. Technically, the OST does not have provisions for undertaking any international review. However, there are some provisions which allow for international consultations. There have been debates in international forums on the need to establish some form of legal doctrine regarding principles governing the conduct of activities in outer space. Nonetheless, efforts made towards establishing a rules-based approach for conduct of activities in outer space have not received any success. Broadly, many have concluded that no mechanism with legally binding provisions is likely to work. At the same, there is a view that although the OST gets more or less adhered to, however, it is not quite serving the purpose for which it was conceived 50 years ago.

The objective of this volume is to understand how this international framework known as the OST has guided the journey of space development. Also, whether it has limited some benign activities. This treaty is heavily biased towards issues concerning the stationing of nuclear weapons in orbit. Obviously, the backdrop for this treaty was the Cold War-era nuclear politics. In the 21st century, the context of space has changed from the Cold War-era narrative, with space weapons becoming a reality. At the same time, the nuclear narrative is refusing to die down. Anti-ballistic missile architecture and related high altitude area defence issues have constrained the emergence of any globally accepted and legally binding mechanism to stall the spread of space-based weapons and anti-satellite weapons. In addition, there are challenges associated with increasing and unregulated space traffic, increasing number of space debris and only a few states showing interest towards developing and testing counter-space technologies. Hence, it is of interest to understand how the OST with its five-decade old 'edifice' is able to address the present and future challenges.

The book has three major sections. The first section is principally shaped by the core concept of this book, i.e. debating the OST mechanism. The second section presents a broad global perspective on the OST, mainly by the spacefaring states. However, it may be noted that these are individual opinions and not official state statements. A fact-finding exercise undertaken through literature surveys, official statements/position papers presented by the states and discussions with experts revealed that various states – both spacefaring and others – have signed the treaty obviously mainly owing to concerns about the misuse of technology, but subsequently fewer deliberations have taken place towards strengthening this mechanism further. Since inherently no review mechanism is available for the OST (like for nuclear, chemical and biological treaty mechanisms after every five years review conferences are organised), hence, there is less debate at the inter- or intra-state level on the OST. The third section of this book largely addresses issues related to governance.

The book is organised around 15 chapters.

The first section lays down the groundwork for the current theoretical environment surrounding the OST. Through the five chapters contained within, it traces the treaty's evolution and trajectory, analysing the modalities and nuances involved in its creation, and its relevance in today's geopolitical framework. In Chapter 1, "The Evolution of the Outer Space Treaty", Ram S. Jakhu provides a preliminary understanding regarding the requirement of a space treaty as it was established in 1963, and the various nuances and caveats involved therein. The chapter attempts to demonstrate the various factors and national perspectives that shaped the specific provisions of the treaty as we understand it today.

Joan Johnson Freese in Chapter 2, "Outer Space Treaty and International Relations Theory: For the Benefit of All Mankind", examines the OST through the lens of international relations (IR) theory. The author reflects that academic considerations regarding the link between IR theory and space have been limited, to say the least, despite increased politicisation of outer space. She attempts to remedy this gap by providing an analysis of the various interpretations of the OST through differing schools of IR thought such as realism, liberalism and constructivism. While the OST attempts to establish the idea of space as a global commons, free from intra-terrestrial conflict, the path towards constructivism and idealised global cooperation is still a space race away.

Continuing with the idea of establishing a theoretical framework, in Chapter 3, "Outer Space Treaty: An Appraisal", G.S. Sachdeva examines the OST with respect to recent developments in the geopolitical balance. The relatively basic framework of the OST has run into challenges against the shifting powerscape of the current multipolar world. He proposes a reappraisal of the OST to address these challenges under three broad segments: the new jurisprudence; original weaknesses of the treaty; and deficiencies due to unvisualised developments that have taken place.

In Chapter 4, "Relevance and Limitations of Outer Space Treaty in the

21st Century", Ranjana Kaul attempts to examine the age-old question of ownership of outer space. Space exploration and the resultant dangers of space colonisation have been linked like the two-headed Janus, and the OST is the only mechanism available at present that deals with placing limitations on spatial exploration. Given the rapid proliferation of space technology in the last few decades, there has been a move away from outer space as a common area free from state sovereignty as the space race has culminated in a move towards 'space superiority'. The author attempts to examine these complexities and developments and the need to contemporise and define the space treaty to make it capable of countering the strategic geospatial challenges of the 21st century.

In Chapter 5, "The Space Arms Race", Ji Yeon Jung makes an attempt to contextualise the shifting geo-political power balance in Asia and the resultant space race between the emergent powers. Through this examination, the author demonstrates increased militarisation of space despite the checks and restraints placed by the OST and various other United Nations resolutions thereafter.

After tracing the evolution of the OST, appraising its various aspects and demonstrating its loopholes, the book attempts to map out the relation between the sovereign interests of individual states and the ways in which they overlap with space exploration. Space has always been an important modality of a state's power projection, and the second section of the book examines the OST as a mechanism that aims to maintain a system of checks and balances in order to prevent a territorialised space war.

In Chapter 6, "The European Union and the Outer Space Treaty: Will the Twain Ever Meet?", Frans G. von der Dunk examines the OST in the context of the European spacescape and the possibility of overlap between the two. The author traces the evolution of the European spacescape and examines its current status vis-à-vis the Galileo and Copernicus projects and then goes on to examine the role of the European Union as a contributor, legislator and space operator with respect to the OST while debating whether the OST and European spacescape will ever meet.

Chapter 7, "50th Anniversary of the Outer Space Treaty: US Contributions as We Move into the 21st Century", continues this deliberation regarding the intersection of national sovereign interests and the space treaty as Philip A. Meek examines the relevance and implications of the OST from the US perspective. The author recapitulates some of the major events that shaped the American space programme into what it is today. He highlights major

US contributions under the OST and current US initiatives which will shape the future of space exploration.

In Chapter 8, "Russia and the Outer Space Treaty", Aleksandr Klapovskiy and Vladimir Yermakov provide an alternate narrative vis-à-vis an examination of the OST through the US's oldest competitor, Russia. They provide a historical context regarding the need for the establishment of such a treaty. The OST emerged in 1967 primarily for the Prevention of an Arms Race in Outer Space (PAROS) and the Prevention of Placement of Weapons (PPW) in outer space. Tracing the historical roots and eventual evolution of the American and Russian space programmes, the authors offer an overview to the various legislations and resolutions that have been passed to control the militarisation of outer space. The success of these initiatives has been mixed, and the several loopholes in the mechanisms need to be addressed and re-examined in the current context. The chapter presents Russia's perspective on the OST and the need for international co-operation to create a sustainable space architecture.

In Chapter 9, "'All's Well that Ends Well': Overview of China's 60-year Space Activity and Space Law", Li Juqian attempts to examine the OST within the context of Chinese national interests. The author summarises the standpoint of China on the overall development of space activities over the past 60 years; discusses the important role of space law treaties in China's legal system and space activities; and observes that China focuses on the peaceful use of outer space and takes an active role in international space law forums to advance the development of legal regime and sustainable space activity.

In Chapter 10, "India and the Outer Space Treaty", Kumar Abhijeet traces the development of India's space programme and its relation with the OST. India has been a party to the OST and has successfully abided by its mandates for the past five decades. Additionally, the various obligations/checks and measures levied by the OST have played an important role in the shaping of India's space law and the various legalities involved. Through an examination of the Indian space legislation in the context of the OST, the author attempts to expose the lacunas in the former and highlights the need to address the shifting paradigms of the latter. Munish Sharma in Chapter 11 discusses the efforts made by both the Koreas and Japan towards implementation of OST at the backdrop of nuclear realities in the region.

A number of international organisations have observer status with COPUOS and they play a significant role in the domain of science technology and law determining the trajectory of outer space activities. One of their most significant contributions has been the OST. In Chapter 12, "50 years of Outer Space Treaty: Reflections of Few Observer Organisations in COPOUS", Malay Adhikari examines how the various observer organisations' activities have followed the general provisions of the OST over the last 50 years and whether trends regarding the same can be identified and isolated.

The third and final section of the book examines the technicalities and legalities of outer space governance and the modifications that need to be made to the OST to make it capable of dealing with concurrent threats and issues. In Chapter 13, "Evolution of Policy and Law for International Space Governance", Eligar Sadeh examines the establishment of policies concerning outer space. The author attempts to provide insight into the norms, practices and compliance mechanisms that are applied to the management of outer space. This is done keeping in mind the emergent challenges of a globalised multipolar world order and the technological complications of the 21st century, including but not limited to orbital debris, militarisation of outer space, observational space sustainability, the idea of space as global commons and globalised cooperation.

Rajeswari Pillai Rajagopalan continues this examination of space sustainability and its various complexities in Chapter 14, "Beyond Outer Space Treaty – Time for New Mechanisms". The author examines the current caveats of the OST with respect to the increasing interdependence being forged between outer space and socio-economic parameters of development. With the interlinking of critical infrastructure and the interstitial overlapping of cyberspace and outer space, the dangers of overpopulation and space debris abound. The author attempts to examine the lacunas in the OST and offer solutions that can be used to build clear definitive frameworks and work towards space sustainability through confidence building measures.

In the final chapter of the book, Chapter 15, titled "The Future of the Outer Space Treaty", Ram S. Jakhu delineates four possible scenarios about the future of the OST and derives crucial implications thereof.

ENDNOTES

1. Euroconsult is the leading global consulting firm specialising in space markets with more than three decades of experience.
2. "20 New Countries to Invest in Space Programs by 2025", Euroconsult, July 19, 2016, at http://euroconsult-ec.com/19_July_2016 (Accessed January 20, 2017).
3. Robert L. Bridge, "International Law and Military Activities in Outer Space", *Akron Law Review*, 13 (4), Spring 1980, p. 658.

4. Rudolf Dolzer, "International Co-operation in Outer Space", *Zeitschrift für ausländisches öffentliches Recht und Völkerrecht*, 45, 1985, pp. 527-546.
5. "Treaty on Principles Governing the Activities of States in the Exploration and Use of Outer Space, including the Moon and Other Celestial Bodies", United Nations Office for Outer Affairs, at http://www.unoosa.org/oosa/en/ourwork/spacelaw/treaties/introouterspacetreaty.html (Accessed December 27, 2015).
6. Rex Zedalis and Catherine Wade, "Anti-Satellite Weapons and the Outer Space Treaty of 1967", *California Western International Law Journal*, 8, 1978, pp. 456-458.
7. Christopher D. Johnson, "Insight - 2017 and the Fiftieth Anniversary of the Outer Space Treaty", Secure World Foundation, January 16, 2017, at https://swfound.org/news/all-news/2017/01/insight-2017-and-the-fiftieth-anniversary-of-the-outer-space-treaty (Accessed January 20, 2017).
8. Christopher D. Johnson, "The Outer Space Treaty at 50", *The Space Review*, January 23, 2017, at http://www.thespacereview.com/article/3155/1 (Accessed January 24, 2017).

SECTION I

DEBATING OUTER SPACE TREATY

1

Evolution of the Outer Space Treaty

Ram S. Jakhu

Introduction

Although the 1967 Outer Space Treaty (OST or Treaty)[1] was the second international agreement, after the adoption of the 1963 Partial Test Ban Treaty,[2] adopted specifically to regulate outer space, it has become the most important and comprehensive international convention governing outer space, celestial bodies and outers pace activities. Some authors call it the constitution of outer space and others equate it with Magna Carta. Though it is neither, the Treaty certainly is foundational in nature and character as the rationale and contents of the four other United Nations (UN) space treaties are essentially based on the provisions of the OST.

This very brief introductory chapter is designed to trace the evolution of the OST with a view to state some relevant facts, factors and national perspectives that shaped the specific provisions of the Treaty. It is believed that the chapter may help the reader of this book to grasp the context of and the spirit behind the letter of various articles and consequently to understand their proper meaning.

The Seeds of the Treaty

The seeds of the Treaty were sown by various political and legal thinkers, mainly from the Western nations, before the start of the space age in 1957. Perhaps the most fundamental and directly relevant views were expressed in

1952 by Oscar Schachter, who was Deputy Director of the UN Legal Department and a recognised authority on international law. He postulated:

> [O]uter space and the celestial bodies would be the common property of all mankind, and no nation would be permitted to exercise domination over any part of it. A legal order would be developed on the principle of free and equal use, with the object of furthering scientific research and investigation. It seems to me that a development of this kind would dramatically emphasize the common heritage of humanity and that it might serve, perhaps significantly, to strengthen the sense of international community which is so vital to the development of a peaceful and secure world order.[3]

Undoubtedly, the views of well-known international legal scholars influenced the thinking of world leaders at the time. However, the physical nature of outer space and the geopolitical atmosphere at the beginning of the space age effectively determined the course of global space governance that ensued and spearheaded political discussions at the UN and conceptualisation of the OST.

The vastness and the physical nature of outer space presented the challenge to nationally own it or to declare exclusive state sovereignty over this new environment. The Soviet Union, which became the first state to 'fly' its space object over the territories of other states, did not face any objection for the violation of national sovereignty of any 'overflown' state – not least the US, which was contemporaneously aspiring to 'fly' over the Soviet territory primarily for reconnaissance purposes. Consequently, the principles of no-sovereignty over outer space and the freedom of exploration and use of outer space emerged. In addition, the UN, particularly through its General Assembly, appeared to be the logical forum for the discussion of newly emerged space-related matters mainly because of the geopolitical necessity for some form of détente between the two antagonists of the Cold War (i.e. the US and Soviet Union). Further, as the embodiment of the international community, which increasingly comprised newly decolonised states, the UN became the guarantor of international goodwill, security and peace.

The UN General Assembly – Mother of the OST

On December 13,1958, the UN General Assembly adopted its first resolution on outer space.[4] The resolution recognised the common interest of mankind in outer space, and that space should be used by all states on the basis of sovereign equality and only for peaceful purposes in order to avoid the extension of the then national rivalries into this new field.[5] In addition, emphasis was placed on the importance of energetically promoting "the fullest exploration and exploitation of outer space for the benefit of mankind".[6] In addition to the recognition of these foundational principles, the UN General Assembly through this resolution took a historic step by establishing an international institution specifically for space affairs (i.e. the then ad hoc Committee on the Peaceful Uses of Outer Space – COPUOS) comprising 18 states selected from all regions of the world representing not only space powers at the time (the US and Soviet Union) but also non-space powers from both the developed and developing countries. The importance of the unique composition of the Committee must not be underestimated because the different perspectives on space matters held by this diverse and representative group of states significantly influenced the nature and scope of the provisions not only of the OST but of other space treaties as well. COPUOS was required to consider and report to the General Assembly, inter alia, on the "nature of legal problems which may arise in the carrying out of programmes to explore outer space".[7]

Geopolitical Realism Sets in

The period between 1959 and 1963 was the most critical and decisive in the evolution of the OST. This was primarily due to the geopolitical situation in the world (height of the Cold War), the emergence of a record number of decolonised nations, and the international community's relentless efforts to charter a brighter future for the humanity as a whole, in particular in the exploration and use of the final frontier. This period not only saw the adoption of basic legal principles to be included in the 1963 Declaration of Legal Principles that would eventually form the foundation of the OST, but it was also during this time that the law-making process for this highly important international treaty for the exploration and use of outer space was cemented.

Despite the initial enthusiasm that accompanied the birth of the newly established ad hoc COPUOS, the Soviet Union was not happy with the composition of the Committee. It believed that the Committee was dominated by the Western nations and their allies. After complex negotiations over a period of two years, the Committee's organisational structure and working procedures were established and have remained largely in place till today.[8] The Committee was established as a permanent Committee of the General Assembly with two Subcommittees, i.e. the Scientific and Technical Subcommittee and Legal Subcommittee, and membership of the Committee was expanded to 24 states. In order to encourage the free and wide exchange of views, it was also decided that summary records, and not detailed minutes of the meetings, were considered to be sufficient. To ensure the views and perspectives of all Committee members be fully heard and respected, it was decided that decisions be made on the basis of consensus. As will be seen in later chapters, the structure and working procedures adopted at the time proved appropriate and conducive for discussing and drafting of the OST, but would present challenges and deadlock as time and space activities develop.

After reiterating the earlier agreed upon principles, eventually in 1961 the UN General Assembly expressed its belief that "the exploration and use of outer space should be only for the betterment of mankind and to the benefit of States irrespective of the stage of their economic or scientific development".[9] In the exploration and use of outer space, states were commended to be guided by the principles that:

(a) International law, including the Charter of the United Nations, applies to outer space and celestial bodies; [and]

(b) Outer space and celestial bodies are free for exploration and use by all States in conformity with international law and are not subject to national appropriation.[10]

Further, in relation to the launch of space objects, states were called upon "to furnish information promptly to the Committee on the Peaceful Uses of Outer Space, through the Secretary-General, for the registration of launchings".[11]

Underlying the fundamental and universally agreed nature of these principles, it should be noted that these principles were, without any significant change, included in the 1963 Declaration of Legal Principles Governing the Activities of States in the Exploration and Use of Outer Space.[12] The contents of this seminal General Assembly resolution formed the foundation on which the most widely-accepted and ratified international agreement governing outer space and space activities would be built.

The Laboured Birth of the OST

The 1963 Declaration of Legal Principles, which was adopted by the UN General Assembly unanimously, was incorporated in toto in the proposals for the draft OST. Undoubtedly, the US and Soviet Union were the main players in the drafting of the Treaty as they actively presented their proposals and counter-proposals. These two space powers did have common interests to protect and grounds to work on, however, often there were serious disagreements on a number of issues. Thus, in the words of Dr. N. Jansentuliayana, who was working for COPUOS and personally witnessed this historic epoch:

> The process of drafting is necessarily detailed, laborious, and time-consuming, involving formal statements of position, general discussions, detailed negotiations, editorial review, and most important, numerous informal consultations which allow delegations to make compromises without having to formally depart from stated positions ... In part this is a consequence of the informal rule that all decisions in the Outer Space Committee and its subsidiary bodies are made by consensus.[13]

It is important to note that non-space faring states, chairmen of the Legal Subcommittee (particularly the Polish legal scholar Manfred Lachs, who later became judge at the International Court of Justice) and COPUOS members from the developing countries played significant, sometime crucial, role in reaching compromise and the finalising the text of the Treaty.[14] The contribution of the developing counties to the formulation of the key provisions of the OST is, unfortunately not well-known and consequently there occur misrepresentations in the proper understanding of the precise meanings of these provisions.[15] For example, during the last days of the discussions on the final draft of the OST, there was a serious deadlock centred on the placement of the following provision:

> The exploration and use of outer space, including the Moon and other celestial bodies, shall be carried out for the benefit and in the interests of all countries, irrespective of their degree of economic or scientific development...

Whereas some argued that this key provision should be incorporated in the operative part of the Treaty, others contended it should be kept in the Treaty's Preamble. The Brazilian proposal was accepted, and favoured the former approach, resulting in the above provision to be legally binding and adopted under Article I(1) of the Treaty. This crucial compromise was hailed, quite appropriately and in unambiguous terms, by the delegations of both the space powers. After the completion of the draft OST in COPUOS, the US delegate stated:

> [the] spirit of compromise shown by the space Powers and the other Powers had produced a treaty which established a *fair balance* between the interests and obligations of all concerned, including the countries which had as yet

> undertaken no space activities... [Article I para. 1] like the provision prohibiting national appropriation by claim of sovereignty, was a *strong safeguard* for those States which at present had no space programme of their own.[16]

Similarly, the Soviet delegate stated that Article I(1), was not "a mere statement of the rights of States", but was designed "to guarantee that the interests, not only of individual States, but of all countries and of the international community as a whole, would be protected".[17] The maintenance of this "fair balance between the interests and obligations of all concerned" is the most fundamental pre-requite not only for the success of the OST but also the global space governance order that it initiated, and as time and space activities develop, would prove indispensable for human progress in outer space.

Acknowledgement: *The author acknowledges with sincere gratitude the help provided by Mr Kuan-Wei (David) Chen in reviewing and providing valuable comments on the earlier draft of this chapter that improved its quality. As always and notwithstanding this invaluable help, the author remains exclusively responsible for any errors contained in this paper: Ram S. Jakhu.*

ENDNOTES

1. *Treaty on Principles Governing the Activities of States in the Exploration and Use of Outer Space, Including the Moon and Other Celestial Bodies,* January 27,1967, 610 UNTS 205, 18 UST 2410, TIAS No 6347, 6 ILM 386 (entered into force on 10 October 1967; as of September 14, 2016, there are 103 ratifications and 25 signatures).
2. *Treaty Banning Nuclear Weapon Tests in the Atmosphere, in Outer Space and Under Water,* August 5, 1963, 480 UNTS 43, 14 UST 1313, TIAS No 5433 (entered into force on October 10, 1963).
3. Oscar Schachter, "Who Owns the Universe?", *Collier's Weekly,* March 22, 1952, 36 at 70.
4. *Question of the Peaceful Use of Outer Space,* GA Res 1348(XIII), UNGAOR, 13th Session, UN Doc A/RES/13/1348(XIII) (1958).
5. Ibid., Preamble, paras. 1-3.
6. Ibid., Preamble, para. 4.
7. Ibid., para. 1(d).
8. *International Co-operation in the Peaceful Uses of Outer Space,* UNGA Res 1472 (XIV), December 12, 1959; *International Co-operation in the Peaceful Uses of Outer Space,* UNGA Res 1721 (XVI), December 20,1961.
9. UNGA Res 1721 (XVI) (A), Preamble, para. 2.
10. Ibid., Preamble, para. 1.
11. UNGA Res 1721 (XVI) (B), para. 1.
12. *Declaration of Legal Principles Concerning the Activities of States in the Exploration and Use of Outer Space,* GA Res 1962 (XVIII), UNGAOR, 18th Session, UN Doc A/RES/18/1962 (1963) [*Declaration of Legal Principles*].

13. N. Jansentuliayana, "The Law-making Process in the United Nations", in N. Jansentuliayana (ed.), *Space Law – Development and Scope*, Praeger, Westport, 1992, p. 33.
14. For an excellent and detailed discussion on the making of the Outer Space Treaty, see Manfred Lachs, *The Law of Outer Space: An Experience in Contemporary Law-Making*, Sijthoff, Leiden, 1972.
15. For details, see Ram S Jakhu, "Developing Countries and the Fundamental Principles of International Space Law", in Wolfgang Abendroth and Rafael Gutiérrez Girardot (eds.), *New Directions in International Law: Essays in Honour of Wolfgang Abendroth: Festschrift zu seinem 75. Geburtstag*, Campus Verlag, Frankfurt, 1982, p. 351.
16. UN General Assembly Official Records (UNGAOR), 21st Session, C1, 1492nd Mtg, UN Doc A/C.1/SR.1492 (1966) at 427-28 [emphasis added].
17. UN Committee on the Peaceful Uses of Outer Space, Legal Subcommittee, 5th Session, 57th Mtg, UN Doc A/AC.105/C.2/SR.57 (1966) at 12.

2

Outer Space Treaty and International Relations Theory: For the Benefit of All Mankind

Joan Johson Freese

Article I of the Outer Space Treaty (OST) states: "The exploration and use of outer space, including the moon and other celestial bodies, shall be carried out for the benefit and in the interests of all countries, irrespective of their degree of economic or scientific development, and shall be the province of all mankind." In other words, outer space is not to be considered by or used only in the interests of individual nation-states. Yet, until very recently, only a few nation-states controlled access to space, owned the most space assets and considered it their domain. While that has been rapidly changing due to lower launch costs and an influx of non-governmental actors, access to and the unfettered use of space is nevertheless considered a "vital national interest" by the US,[1] a revitalised national interest by Russia and an aspiring national interest by China, India and many other countries. Many forms and aspects of national security are seen to flow from and through space. Consequently, terrestrial geopolitics inherently affects national interpretations of not just Article I, but the entire OST.

Theories of international relations (IR) are essentially mental maps, or lens, providing individuals, often decision-makers, views of the world. Academic considerations of the link between IR theories and space have been limited. In fact, space has been largely "removed, marginalised and silenced"[2] from the discourse that generally pervades IR, and instead subjected to largely action-reaction and inertia-driven policies and strategies. Part of the problem is a failure, or disinterest, among some technology-oriented space players to understand the basic premises of an IR theory. For more policy-oriented individuals, the problem can be in failing to consider how IR theories relate to the unique nature and challenges of space. And, among both, there is sometimes a tendency to confuse strategic means and ends used to further policies adopted within IR frameworks.

Realism posits that the world is a dangerous, competitive place where nation-states are all important, power is zero-sum, and self-interest and self-protection are key. Under realism, alliances are constantly shifting according to self-interest. Liberal-internationalism, on the other hand, presents the world as fundamentally more complex, with key actors extending beyond nation-states, winning and losing not necessarily zero-sum, and norms and international institutions offering long-term solution sets to problems. Conflict is still an option, sometimes even a choice to, for example, spread liberal internationalism, but not a first option. Constructivism, the most abstract of the primary IR theories, sees the world as run by ideas. The how and why of the wide-spread changing of ideas and having ideas become accepted is key to the constructivists' view of the world.

Analysts and even some politicians have suggested that in terms of space security, there are two basic camps: Realists who understand that inherent competition in space will inevitably lead to conflict,[3] and liberal-internationalists who seek to use arms control – including legal means such as the OST – to avoid conflict. However, accepting conflict as inevitable negates the perceived value of arms control efforts; indeed, arms control, it is suggested, could place even the most-advanced countries at a disadvantage by holding them back from further advancement while others catch up. Moreover, if operating from a zero-sum perspective, seeking to advance benefits for all mankind can be interpreted as doing so at the expense of individual nation-states. Hence realism – and inevitable competition and so the need to prepare for it – has often prevailed as the view of nation-states that consider space as, or potentially as, a "vital national interest".

In terms of strategy, however, power as a goal and power as a means to achieve a goal are very different and should not be conflated. Realism dictates that nation-states will support their own self-interests, and there are many varieties of power that can be used to achieve those self-interests. Therefore, achieving ends – goals – can be considered the essential element of realism, not the means used to get there. Because of the unique nature of space as a domain, the OST, while affording liberal-internationalist benefits to all mankind, also provides valuable means for the goals of realist nation-states – access to and use of space – to be achieved. Means most often associated with liberal-internationalism – rules, norms and international cooperation to manage use of the environment – may in fact be the best option for achieving space-related goals considered as vital national interests.

Outer space is a global commons. Valid comparisons have been made with other "common" domains, but those comparisons have limitations. In terms of both the air and sea, for example, while portions of each are considered beyond the jurisdiction of any one nation-state, nation-states do have legal jurisdiction over portions near or above their territories. The same is not largely true for outer space.[4] In security terms, while air dominance can be established by one country for a limited period over a limited space with decision-makers having relatively full situational awareness of the area during that period, the tyranny of distance (among other factors) means the same cannot be said of outer space. But it is a commons in the sense that the damage or destruction of outer space environment by one will result in an inability to use the environment for all. Therefore, it is in the self-interest of all countries that consider outer space a "vital interest" to do all they can to preserve the environment. The OST has provided the foundation for preserving the outer space environment for the past 40 years. Article IV in particular is an example of how that is the case.

It is within Article IV that the key arms control provisions reside. States party to the Treaty agree (1) not to place in orbit around the Earth any objects carrying nuclear weapons or any other kinds of weapons of mass destruction, install such weapons on celestial bodies, or station such weapons in outer space in any other manner; (2) that the Moon and other celestial bodies shall be used exclusively for peaceful purposes; (3) that the establishment of military bases, installations and fortifications, the testing of any type of weapons and the conduct of military manoeuvres on celestial bodies shall be forbidden. However, the use of military personnel for scientific research or for any other peaceful purposes is not prohibited, nor is the use of any equipment or facility necessary for peaceful exploration of the Moon and other celestial bodies.

While weapons of mass destruction are prohibited in space, weapons in general are not. Experience has shown that the release of energy or use of kinetic force in space – weapons, regardless of the type or whether for offensive or defensive purposes – can pollute the environment, making it more expensive and dangerous to use, limiting its viability as a domain and ultimately threatening the sustainability of the environment. From the 1962 US Starfish Prime test of nuclear weapons in space, to the Chinese anti-satellite weapons test (ASAT) in 2007, electro-magnetic pulse and/or debris generating events have been shown to create environmental damage requiring years for recovery. Indeed all debris-creating events, regardless of origin, are increasingly recognised by space-faring nations as the greatest risk to the space environment, and requiring both restraint and cooperation among nations for the issue to be effectively addressed. There are a multitude of issues emanating from acknowledgement of the space debris issue, none of which require a military solution, but all requiring cooperation and management.

The OST offered parameters for space activities appropriate for the time and technology of the 1960s. Those parameters are still appropriate some 40 years later, but there is a need for supplementary rules, norms and international laws to catch up to technology.

While it would be nice if nation-states all worked together because it is "the right thing to do", but that rarely happens. Most often, nation-states work together because it is in their best interest to do so. Sometimes cooperation serves individual as well as the collective interests of these nation-states. Such is the case with outer space: Realist goals are best achieved by liberal-internationalist means. Perhaps some day the idea of intra-terrestrial competition will change, and constructivism will prevail, though that will most likely only be if there is a different, greater competition facing mankind. In the meantime, however, as long as terrestrial geopolitics is characterised as competitive, and space is considered "congested, contested and competitive", self-interest will rule. Consequently, it is critical that nation-states differentiate and align strategic means and ends, rather than making their own circumstances worse.

ENDNOTES

1. Robert G. Joseph, Undersecretary of State for Arms Control and International Security, "Remarks to Center for Space and Defense Forum", January 11, 2007, at http://2001-2009.state.gov/t/us/rm/78679.htm.
2. Natalie Bormann and Michael Sheehan (eds.), "Introduction", *Securing Outer Space*, Routledge, London/New York, 2009.
3. Further evidence for this view is drawn from the assertion that all other domains have become subject to warfare; therefore, the weaponisation of space is inevitable.
4. Tonga tried to claim the rights to 16 valuable slots in geosynchronous orbit directly over the country. Edmund L. Andrews, "Tiny Tonga Seeks Satellite Empire in Space", *The New York Times*, August 28, 1990, at http://www.nytimes.com/1990/08/28/business/tiny-tonga-seeks-satellite-empire-in-space.html?pagewanted=all.

3

Outer Space Treaty: An Appraisal

G.S. Sachdeva

Introduction

The Outer Space Treaty (OST)[1] has been in operation for 50 years. It has proved its mettle and resilience to ensuing changes and advancing technologies to a great extent and for a long time. The drafters deserve commendation for their ingenuity in creating a farsighted and durable instrument. However, lately the Treaty has been under pressure. Clear mandates and straight answers have not been available to certain developments and contingencies. As a result, additional convention, agreements and soft law principles and guidelines have been resorted to and adopted. Some questions still remain unanswered, and gaps in its legal edifice are discernible which need to be specified and filled.

The Treaty has run into challenges for various reasons. First, because certain contingencies could not be visualised and catered for at its inception. Second, for some deficiencies compromises were necessary to draw consensus among negotiating members. It is now nearing a critical point of adjustments. The deficiencies thus constitute a motley mix in the face of unforeseen developments and deserve to be reviewed and rectified to make space law direct, explicit and unequivocal to retain its relevance. Overall, the OST must be respected because in outer space failure of governance is not an option. Hence, this crisis of credibility in space law needs bridges of cooperation, confidence and trust.

The OST is a basic document that is superb in content and contention, embodying laudable concepts and settled legal precepts. It is a political document of stupendous legal import and forms an over-arching oeuvre in the context of space law. It is often euphemistically called the *Grundnorm* of space law and a mascot of space governance. It was a document containing new jurisprudence and novel legal concepts when created, but has been overtaken by rapid spread of the footprint of technology and overtures of private enterprise. Therefore, over the last two decades of its operation, it has started showing fault lines which should not be trivialised but remedied holistically to make it contemporary and to retain its relevance to the future.

In a scenario where space technology is literally galloping, business possibilities are exfoliating, space manufacturing is showing signs of viability, the reach of space law is becoming so expansive and applications so varied that the existing corpus of space law is far too inadequate for its object and purpose. The Treaty Law that evolved in the first two decades of the initial space odysseys has been singularly state-centric and not been prescient enough about scientific development, or the clout of private enterprise to overtake public spending in space activities or the consequences of space activities like accumulation of space debris. The new scenario with a wide vista of exploration, commercial viabilities and exploitative uses needs an amplified, pervasive and binding legal regimen.

Having identified some lacunae in the extant Treaty, it transpires that the OST needs to be overhauled and updated in order to accommodate existing adjuncts, adjust to emerging possibilities, incorporate new visions on the technology threshold and accommodate widening business horizon. These challenges are upon us and stare into our face. We can shut our eyes only at our own peril and risk to survival of humanity. Thus, there is need for a holistic and visionary approach to revise the OST where strategic priorities and national interests do not interfere in the New Space Law for New Space Vista.

It is, therefore, proposed to appraise the OST under three headings: the new jurisprudence; original weaknesses of the Treaty; and deficiencies due to unvisualised developments that have taken place. Based on the appraisal, solutions would be explored to ameliorate the legal milieu and contemporise the OST. Obviously, under the circumstances no binary answers can be offered. Further, conscious of the fact that negotiation of treaties is laborious and time-consuming, and that minor tinkering would be neither adequate nor futuristic enough, this chapter proposes initiation of specialised protocol to amplify germane law to overcome each weakness directly and specifically.

Moreover, the chapter also mentions the likely future inadequacies of the Office of Space Affairs (OOSA) due to additional burden of monitoring tasks, and it is advocated that the OOSA be upgraded to an organ of the United Nations (UN), such as World Space Organisation (WSO), that acts as a Trustee of outer space and is duly empowered, professionally competent and equal to the future tasks and responsibilities of space governance, possible financial accruals from launches, enforcement of compliances under law and redressal of disputes in the first instance.

The New Jurisprudence

It is creditworthy that framers of the OST have really been innovative in their conceptual thinking and legal acumen to introduce new tenets of jurisprudence that are different from terrestrial laws and settled principles of international law. The new concepts are specific to the domain of outer space and celestial bodies other than the Earth. These principles have shown their merit and constitute the New Jurisprudence. These are discussed in succeeding paragraphs.

Principle of Non-Appropriation

Humankind has traditionally espoused sovereignty and has lived with state boundaries from times immemorial. Annexation, occupation and appropriation are legal acts under terrestrial laws. Therefore, the provision of the OST embodying the principle of community ownership of outer space and celestial bodies without frontiers appears novel and innovative. Indeed, a great departure from the age-old mindset of winning over by war or use of colonial territory and assuming sovereign status. It has thus obviated a reckless race for contentious occupation to demark boundaries in outer space. In a way, the Treaty mandates that outer space and celestial bodies are to be regarded as *res nullius* or better still *res communis* or *res publica*.

This is therefore, an explicit and prescient articulation in the Treaty that prohibits national sovereignty in outer space. The text of the Treaty reads, "Outer space including the Moon and other celestial bodies, is not subject to national appropriation by claim of sovereignty, by means of use or occupation, or by any other means."[2] This clearly means that there will be no frontiers in outer space and on celestial bodies. It thus constitutes a quantum leap into a new and a higher level of legal regimen that has now been accepted universally to become an article of faith for international community and its breaches are intolerable.

In other words, this principle reveals a fundamental difference from terrestrial legal regimes of property ownership and state sovereignty, whereas the Treaty asserts the principles of non-appropriation of spatial property and non-sovereignty over celestial real estate. The Treaty prohibits all states of international comity from declaring national ownership in any manner or for any type of proprietary rights in the outer space or on any celestial body for any reason or means or by any method. This has saved the human race a scramble for celestial occupation or ownership.

Despite such lucid and firm provision in the OST, the US is creating an invalid exception through its national legislation called "Apollo Lunar Landing Legacy Bill".[3] This Bill proposes to establish National Historical Park on the Moon over areas where Apollo landings took place in 1969 and 1972; and some items and accoutrements of the Mission still remain scattered there. It is ostensibly intended to protect and preserve history for posterity and shall be administered by the US Government. The idea may be good and authenticity of site may also not be in dispute, yet the act of acquisition of the area and siting of a permanent Park on the Moon 'under domestic control' fouls with the established space law.[4] And the US has no jurisdiction or *locus standi* to legislate on the real estate of the celestial bodies and unilaterally appropriate control over a part of it.

The Bill, of course, has expired,[5] but perhaps it was intended only to test waters and gauge resistance to the proposal. Regrettably, the world comity has not openly come out in condemnation of such a blatant initiative towards covert colonisation of the Moon; and subsequently, may be other celestial bodies. The response is rather muted and only from isolated quarters.[6] Even Russia, China and India have vocalised no official stand on this contentious issue. Few have questioned the very utility for such a park. And if, at all, justifiably needed, it can be placed under control and jurisdiction of the UN administration in some manner or expedient modus.

Province of All Mankind

The Treaty, in true altruism, recognises "the common interest of all mankind"[7] in outer space. This shows the rights, concern and responsibility of the entire comity of nations and not of spacefaring or treaty-ratifying nations alone. Treaty further asserts that "outer space including the Moon and other celestial bodies shall be the province of all mankind"[8] implying that the entire universe belongs to humanity as a whole and is thus *res communis*. A variant of this is the principle of Common Heritage of Mankind. The meaning of this clause is not explicit and may lead to doubts, but the intent and spirit of this accepts the interest and virtual control of entire mankind and not only of the Treaty member-states.

In general connotation, the conceptual phrase "province of mankind" has two parts: first, the word 'province' emphasises historical distinctness and character differentiation of the expanse of outer space and recognises discrete territorial traits of the celestial bodies that comprise the 'space system' – a new frontier that is strategically and politically divergent from the planet earth. In other words, the entire universe minus the planet earth is the province of mankind. Second, the word 'mankind' implies humanity. But both terms are generic and vague in what they encompass today and in the future. Further, it is arguable that these have no legal entity or legal capacity as either subject or object of international law. In the Treaty, the concept of 'province of mankind' is not explicated and possibly does not relate to legality but has been stated as a precept of unity and altruism implying that activities in the outer space "shall be carried out for the benefit and in the interest of all countries…".[9] It is thus a primary principle of space jurisprudence with humanitarian nuances and a tinge of globally 'communitised' resources.

Outer space "shall be the province of all mankind"[10] shifts the emphasis from the traditional postulate of national sovereignty to international cooperation, with community rights for common good, thus highlighting the underlying principle that there are areas where common interests of mankind must be served and given primacy. This clause concedes the possibility of conflict of ideology or clash of national interests in space operations, but dispels "any such spectre to seek a common vision of their future relations in a newly accessible environment".[11] This principle strengthens the sense of international community with de facto respect to other countries to create common interest and encourage collective security for the sake of mankind.

It thus brings in sharp relief the concept of community ownership of outer space and celestial bodies as a 'province of mankind'. This is in contrast to the rules of territoriality under international law on the planet earth. But no dissent has been vocalised to challenge this norm of space law. Outer space thus remains *res* of mankind, for use by mankind and for the ultimate welfare and benefit of mankind. The wisdom of this postulate has empirically proven itself and the unanimity in its acceptance by the states elevates it to become a primary principle of space jurisprudence.

Freedom of Access to All States

As a corollary to the province of mankind, there is another innovative clause of liberated thinking that grants a right of access to outer space to all states globally and does not restrict this freedom to the member-states only as per the usual practice of states in international treaty relations. The egalitarian approach and universality in practice is indeed commendable in context and spirit. And freedom implies no necessity for prior permission from any authority or organisation to take on space activities. It is, indeed, a great step forward towards universalisation of the Treaty and in the direction of New Jurisprudence of outer space.

The Treaty provision states, "[O]uter space, including the Moon and other celestial bodies, shall be free for exploration and use by all States without discrimination of any kind on basis of equality and in accordance with international law, and there shall be free access to all areas of celestial bodies [emphasis added]."[12] This provision has a novel inclusiveness to assure that no state suffers discrimination or is disadvantaged by law for reason of not being member of the Treaty. Thus, this Article has two dimensions, i.e. freedom of use by all states and free access to all areas of celestial bodies. However liberal be the guarantee or strictness of its assurance, it still devolves certain reasonable restrictions to assure equal right of enjoyment of the same by the other countries.

Thus, this freedom is not absolute and carries with it certain riders and corresponding duties. These are in mutual and reciprocal interest to avoid interference in each other's activities and interests. The Treaty mandates that "...the states shall be guided by the principle of cooperation and mutual assistance and shall conduct all their activities in outer space, including the Moon and other celestial bodies with due regard to the corresponding interests of all other States Parties to the Treaty".[13] This egalitarian and altruistic provision sounds basic and fundamental yet could have distinct and derogatory nuances. For example, it may create complications on international liability where terrorists procure a launch from private consortium with mal-intent, or if launching state is not a member of the OST or launch is procured by a non-OST member state. The situation can be made more complex by factoring in more imponderables.

There is another assurance of additional "freedom of scientific investigation in outer space, including the moon and other celestial bodies, and states shall facilitate and encourage international cooperation in such investigation".[14] This facilitation is also circumscribed by correlative duties and individual obligations. The states shall undertake experimentation and "...pursue studies of outer space, including the moon and other celestial bodies, and conduct exploration of them so as to avoid harmful contamination and also adverse changes in the environment of the earth resulting from the introduction of the extra-terrestrial matter and, where necessary, shall adopt appropriate measures for this purpose".[15] The freedom is certain yet the duties devolving are definite and distinct.

In Benefit and Interest of All Countries

Normally, pacts and treaties protect and ensure interests of participating states that are bound by mutual obligations of the *pacta.* Thus, treaty members constitute an exclusive group that shares beneficial accruals and keeps other states disfavoured of the benefits arising from the treaty by embargoes and differentials. But the OST is from a different genre and assures benefits to all states and espouses inclusivity and 'globality'. The altruistic element pervades through the Treaty.

The Treaty provides that, "[T]he exploration and use of outer space, including the Moon and other celestial bodies, shall be carried out for the benefit and in the interest of all countries, irrespective of their degree of economic or scientific development."[16] Here again the expressed desire of sharing benefits and results of scientific investigation is laudable on the principle of equality and in uplifting of the developing states, economically, socially and on technology. This makes for an innovative clause in the Treaty for sharing of benefits even with non-participating, non-state-faring states.

Humanity has never seen such humanitarian, egalitarian and universalised treatment or character of freedom under any instrument of international law. This laudable idea needs to be emulated and incorporated in other future treaties. It ushers a new era of goodwill towards all and introduces a novel quality of New Jurisprudence in space relations. It makes for a commendable trend in international law in general.

International Cooperation as Cardinal Principle

Cooperation is implied in every pact, but in this Treaty, international cooperation has been urged and exhorted at every step. The OST is replete with references to cooperation between states, and the relevant provisions are meaningful and binding. The principle of international cooperation runs as a basic thread throughout the Treaty as a cohesive force urging states to cooperate through facilitation and consultation in their mutual benefit, overall interest of humankind, common scientific pursuits, global peace and security and above all, humanitarian considerations.

To begin with the references, the Preamble to the Treaty optimistically exhorts state parties "to contribute to broad international cooperation in the scientific as well as the legal aspects of the exploration and use of outer space for peaceful purposes". It further believes "...that such cooperation will contribute to the development of mutual understanding and to the strengthening of friendly relations between states and peoples".[17] The hope is sincere, and the good intention permeates all through the Treaty.

Article I of the OST mentions while referring to the freedom of scientific investigation in outer space and celestial bodies that, "... [S]tates shall facilitate and encourage international cooperation in such investigation." Again Article III while permitting space activities urges for "...maintaining international peace and security and promoting international cooperation and understanding". Further, Article X endorses international cooperation to afford "...an opportunity to observe the flight of space objects launched by ... states". Extending the principle, Article XI commands to "promote international cooperation in the peaceful exploration and use of outer space...including the moon and other celestial bodies...to inform...of the nature, conduct, locations and results of such activities".[18] There are other instruments, too, propagating this philosophy, like Declaration of International Cooperation, 1996.

One can also solicit support from the views of Goedhuis that in meeting the varied challenges of the space age man has been able to cavort and combine the forces of the social complex which provide a realisation of greater world interdependence because of limitations of technology and ultra-hazardous domain of outer space and thus necessitate cooperation. Inclusivity of all states, technically capable or still struggling is inalienable and integral to the order of outer space. Thus, an important feature of Space Law reflects the gradually transforming structure and reveals a process to detoxify international relations of the phantom of sovereignty and highlights recognition of the compulsion of international cooperation in the field of outer space.[19]

It is thus held by many scholars that Space Law contains stronger cooperative duties and obligation to collaboration than general international law. Rüdiger Wolfrum has particularly stressed that this principle marks a significant break and a conceptual transition away from the traditional international law of co-existence to a new law of cooperation.[20] Similarly, Rudolf Dolzer holds the view that the structure of Space Law is based on active cooperation and mutual assistance complemented by specialised duties towards activities in and relating to outer space. This reflects a concept of obligation of assistance with voluntary spontaneity and in the spirit of reciprocity.[21]

From the foregoing analysis, it gets amply substantiated that international cooperation is the cornerstone of space law for safer space activities, betterment of mankind and improvement in quality of life on the earth. International cooperation is not a peripheral issue but a strong strand that runs through the regime of space law and becomes common denominator of treaties and agreements, principles and guidelines. Due to its importance, this concept has been fully imbibed and internalised in the OST and in the general corpus of space law; and has thus become integral to its functioning as well as normative behaviour of all states. Thus, this principle undoubtedly assumes the cardinal status and introduces New Jurisprudence of peace and security, eschewing conflict and confrontation.

Original Weaknesses in the Treaty

There are some original deficiencies in the OST which have existed *ab initio* due to certain reasons or compulsions. These, therefore, detract from the value and merit of the Treaty and leave chinks in its efficacious compliance. These intrinsic weaknesses of the Treaty make it susceptible to vested interpretation. A few illustrative deficiencies of this nature are discussed in succeeding paragraphs.

Absence of Definition of Important Terms Used

Normally, definitions of important terms form the core of a treaty because the entire meaning and interpretation of the treaty clauses are dependent on them. It is common knowledge in diplomatic circles that this is the most difficult part of negotiations, yet their necessity cannot be underrated. It, nevertheless, remains most vital to its core coverage and for governance of the regime. It is axiomatic that even the best of legal statutes can be defeated in their intent and purpose due to sheer lack of definitions of operative words or for any ambiguity therein. The same delinquency prevails in the Space Treaty that makes it susceptible to equivocal interpretation and, at times, untenable construction of its provision resulting in foiling the intended motives of the Treaty. For example, the US legislation to establish a Heritage Park on the Moon and the other eliciting Competition from Private Enterprise to harvest mineral resources from asteroid.[22]

Such faultlines have always existed in the OST and have been its bane.[23] Definition of important terms used in the Treaty like astronaut, launch vehicle, peaceful purpose, placement in earth orbit, space object, space debris, weapons of mass destruction (WMD), province of mankind, etc. could have found place for good utility. Undue promiscuity in legal interpretation, in an utter disregard of the very intent and objects of the treaty, is indeed reprehensible, yet state parties do get tempted in their vested interests and own advancement

Legal experts have long wrestled with the true meaning and actual intent of some of the words contained in the Treaty, like legitimisation of private enterprise in space activities or justification of Anti-Ballistic Missiles in outer space. Erudition of the scholars is commendable, but they tend to forget their professional burden implied in the maxim, *ex vinculus sermocinatur*. This enjoins that treaty must be interpreted in good faith and in its ordinary meaning. This duty is also cast under the Vienna Convention on the Law of Treaties, 1969. We can ignore this prime principle only by betrayal of our conscience. Nevertheless, definitions in a treaty assume primary importance in its implementation towards conflict mitigation, international security and world peace.

This deficiency has led to problems in *ad idem* understanding by the parties; and construing true and proper meaning of the same term. Differences have arisen even among space scholars who have interpreted the same term in the hue of national interest. The latest examples are the anti-satellite (ASAT) tests undertaken by China and later by the US to kill their own satellites in orbit in a flagrant display of technological superiority and causing considerable space debris with impunity. These actions were certainly not peaceful activities, nor could be squarely dubbed as military missions though the responsibility of causing unnecessary and artificial debris attaches to both.

Further, some phrases appear political rhetoric or sheer platitudes. An example could be 'province of mankind', which hardly denotes any clear or unambiguous concept with meaning or precedence in law. It is a laudable clause but with little legal import or relevance. Another example can be "Astronauts are Envoys of Mankind in outer space'. The halo of ambassadorial status is meaningless without other formalities and credentials. First of all, who is an astronaut? Moreover, every astronaut in a situation of multiples like in the International Space Station (ISS) cannot be an envoy of mankind.[24] Therefore, correctly and lucidly defining the important terms in a treaty becomes an important and unavoidable task.

Inadequate Provision Relating to Prohibition on Bombs

This Article is a half-hearted drafting and its deficiency can be surmised to the compulsion of achieving consensus in negotiations. There is no gainsaying the fact that, historically, activities into the outer space were an off-shoot of the arms race and a corollary to the development of missile defence projects by the super powers. These activities were controlled and operated under military domain to exploit the outer space for their national defence imperatives or security cover for their strategic allies. Therefore, the risk of weaponisation of outer space was posed by only a couple of countries. The framers of the OST were naturally concerned about the possibilities arising from such a scenario, particularly the use of outer space for military activities. Thus, the germane Article, which is barely suitable, yet adequate enough to gain consensus, was embodied in the OST.

The Treaty provision requires "State-Parties to undertake not to place in orbit around the earth any objects carrying nuclear weapons or any other kinds of weapons of mass destruction, install such weapons on celestial bodies, or station weapons in outer space in any other manner".[25] The Treaty further ordains that "the Moon and other celestial bodies shall be used…exclusively for peaceful purposes". This inadequate provision was supplemented and reinforced by the Anti-Ballistic Missile Treaty, 1972, between the US and USSR. This accepted a voluntary moratorium on the development and testing of anti-ballistic missiles. The agreement placed a limit on deployment of anti-ballistic missiles at status quo. Ostensible *ratio* was because these missiles operate through the jurisdiction of outer space. However, in 2002 the US reneged on this Agreement.

Later, when the ICBMs came of age, and the debate on Star Wars was in heat with the possibility of Soviet Fractional Orbital Bombing System (FOBS)[26] becoming functional, the SALT-II Treaty was signed in 1979. This specifically provided, "Each party undertakes not to develop, test or deploy...systems for placing into earth orbit nuclear weapons or any other kind of weapons of mass destruction, including fractional orbital missiles." This had assured the world that mutually assured destruction (MAD) had been averted for the time being. The SALT-II Treaty endorsed what was already there in the 1967 OST.[27] But ominous threats still exist from the BMD Initiative of the US, mooted in 2012, and Chinese development of Lasers, developed in 2006, that can blind the missiles.

The obvious intention of the OST was to exclude bombs and military manoeuvres from outer space to ensure safety, security and collective survival of humanity but to achieve this, it introduced a lame clause. Its deficiency comes to fore in relation to conventional kinetic, directed energy weapons (DEW) and the so-called below-WMD level bombs. It also does not specifically ban development, testing and deployment of any lower kind of weapons. The endemic weakness is amply vouchsafed by other agreements in tandem which extend the ambit of its scope and reinforce the purpose.

The OST can, however, be further improved in this regard, in content and efficacy, by adding that "no space object or launched space vehicle or weapon of any kind that enters the domain of outer space shall target or hit any point on the planet earth".[28] The US and former Soviet Union had already agreed to a similar provision in the Anti-Ballistic Missile Treaty. Thus, the merit of the proposed clause stands vindicated, which should make its negotiation easier.

Astronauts: Envoys of Mankind in Outer Space

The provision of the OST that includes astronauts as the envoys of mankind, though highly sensitive and innovative, seems to have been inserted as political rhetoric because the clause is simply incomplete and inadequate for the intent expressed and legalities involved. Possibly, such a passionate phrase of ambassadorial status has been used to appeal to all the states to accord due diplomatic immunities to protect astronauts and solicit their unhesitating cooperation for rescue and safety. It was laudable to accord a halo of a hero to honour the daring adventure of an astronaut. It was a logical requirement but was used as a valuable ruse for eliciting cooperation during the Cold War period. Now, however, with Agreement on Rescue and Return of Astronauts, etc., 1968 in place, it makes no diplomatic sense and still lesser legal sense. This clause seems to have already served its purpose and outlived its utility. In fact, the wisdom of this dictum has been questioned and legal basis of this status has been controverted.[29]

To put it into perspective, the OST under Article V regards astronauts as "envoys of mankind" in outer space but offers no definitional statement to identify such a persona. Definitions of the term 'astronaut' by different countries are rather loose and non-descript. Take the US definition, for example, it is so open-ended that it includes even those trained for space journey as astronauts.[30] Other countries have their own terminology and lexicon. Soviets call astronauts as Cosmonauts and Chinese, Taikonauts. Thus, different countries accord different definitions which are neither congruent nor compatible.

It is ironical that after the OST, none of the treaties or agreements has used the phrase 'envoy of mankind'. A comparative analysis of the OST and other related international instruments including Agreement on Rescue and Return of Astronauts reveals an extensive use of the term 'astronaut'. But this term has variously been alluded to for indicating different types of space farers who have in several texts and contexts been referred to and addressed differently. For example, astronaut is used in Article 5 of the OST and Preamble to the Rescue Agreement; personnel in Article 8 of the OST and Articles 1 and 4 of the Rescue Agreement; representative in Article 12 of the OST; persons on board a space object in Articles 3 and 4 of the Convention on Liability; and crew in Article 8 of the Russian text of the OST). The references, titles and substitute terms used for astronauts evade commonality and *ad idem* understanding.

The basic problem is that no definition of 'astronaut' has been enshrined in any treaty or agreement which makes legal position of multiple astronauts at ISS, all being envoys of mankind at the same time and place, rather paradoxical. Even procedural and credential requirements are not laid down anywhere. Be that as it may, it would appear sagacious to designate only the commander of the space vehicle to act as Envoy of Mankind, if at all such a provision is to be retained and respected. The dilemma is serious but has existed *abinitio*. This deficiency needs to be corrected.

Deficiencies Due to Unenvisioned Developments

There are certain deficiencies in the OST that are due to developments which could not be visualised at the time of drafting of the Treaty. These have come about later due to different reasons and impetus. Nevertheless, these have overtaken the Treaty in content and scope. These developments relate to quick advancement in space technology, the growth of private enterprise to undertake space activities on their own and competitively, and lastly the exponential increase in the quantum of space debris accumulating in outer space. These are only illustrations of the deficiencies.

Rapid Advancement of Space Technology

Development of technology has been a progressive process occurring in multi-dimensions but its acceleration in the late years of the 20th century was phenomenal indeed. The beginning of space activities for many countries was a spin-off from the military rocket technologies being developed by the super powers for strategic objectives and military domination. Adaptation to space activities was an easy next step and natural advancement. And it did occur as a logical transformation in the 1950s.

Space technology and its applications have advanced at an unimaginable and unprecedented speed over last half a century. Its progression from Sputnik to Space Shuttles and ISS and to Juno reaching the Jupiter has been amazing in scientific history. The Juno deep-space probe travelled for five years braving intense radiation to reach its destination on July 4, 2016 to explore the king of planets.[31] Earlier probes were the National Aeronautics and Space Administration (NASA) Mariner and India's *Mangalyan* to the Mars, among others. The Moon has been probed by many countries and its colonisation[32] is on the cards.

Next, coming to space applications, the early ones related to national strategy and technological domination, but that was in the era of two super powers – the US and USSR. Today, the countries are playing a different ball game. The US is coaxing private enterprises to compete for mining of asteroids through domestic legislation.[33] Other spacefaring countries have mostly developed applications for national socio-economic development and peaceful purposes. For example, India has harnessed space technology to civilian uses for rural uplift, literacy and education of masses and for social reach like tele-medicine and information dissemination. Among national projects are those extending the footprint of broadcasting and communications, weather forecasting, disaster management and high-quality remote-sensing facilities.

Space innovation is intended to reduce the digital divide and make space-based utilities accessible and affordable to all countries. The motto should be to assist humanity to absorb benefits of space assets, maximise efficiency to make human life comfortable with utmost transparency and at the same time protecting privacy. The contemporary threshold of technology unfolded by time was not imaginable at the point of adoption of the Treaty; and today, this motto cannot be achieved by voluntary self-regulation by spacefaring countries for reason of controls on technology transfer or due to domestic compulsions – political, legal or financial. Hence, there is need for a remedial and quasi-penal regime that can be enforced by appropriate mechanisms or through an empowered UN Institution.

Enthusiasm of Private Enterprise – Exploration to Exploitation

Considering that space entities of two super powers were the only actors in outer space in the 1950s; and assuming their monopoly in space activities in the foreseeable future for reasons of technology and financial investments, the OST was drafted as a state-centric document. It thus treated state organisations and controlled entities as the only possible players in outer space. Hence, the focus of space law was mainly on state as unit of activity and control. It was then hardly imaginable that private entrepreneurs would be able to muster courage, possess funds and harness diverse technologies to undertake space activities.

It was then believed that space activities involved advanced technology needing high level of research and development backup, coordinating a wide array of augmenting technologies, long gestation period of projects, hazardous space environment, endemic risks in operations and, above all, huge capital outlays and associated financial imponderables. All these building blocks had their own peculiar challenges individually and incongruence in combination. No wonder, the situation seemed insurmountable except with the positive support of governmental departments acting in concert in the public interest. The hurdles seemed too unpredictable and almost insurmountable for profit-oriented commercial enterprise.

Be that as it may, private enterprise has, within half a century, overcome all the apprehensions and made its commercial calculations. Business houses, e.g. Space X, Virgin Galactic, Space-Ship One and Bigelow Aerospace, all have found overall competence to commence operations, and that too, to successfully perform space duties for the state agencies, like NASA, in operating shuttles to carry cargo to the ISS and return wastes to the earth. This effort is indeed commendable and deserves full impetus while manned space travel and space hospitality are just waiting in the wings. But at the same time, the private sector should not forget its corporate social responsibility (CSR). Thus, there is need for compassionate capitalism and value-based business ethics in space commerce and industry.

The outer space arena is thus getting crowded with all kinds of actors: state entities, private players and business consortia, each with different motivations, priorities and concerns that are likely to cause conflict of interests. This situation, however, confronts a constraint in the OST that does not directly and explicitly permit activities of private actors in outer space. The only oblique reference is found in Article VI of the OST dealing with state responsibility. It mandates that State "shall bear international responsibility for national activities in outer space...whether such activities are carried on by governmental agencies or by non-governmental entities". The latter, however, "shall require authorisation and continuing supervision by the appropriate State Party...". A liberal or permissive interpretation of this Article may include private enterprise, but a specific mention is missing.

Today, when private players are vying to be on the centre stage of the space arena, it seems advisable to willingly accept them in the fold of space actors and accord them due recognition and commensurate importance. The incorporation of this development may need only a minor addition in the Treaty. Or pro-actively considering that they may enter other areas of space activities like mining of asteroids or extraction of minerals from celestial bodies or the competition to space travel and space tourism, it may seem prudent to draft a specific protocol that is futuristic enough to cover and provide regulations for prime operations and ancillary aspects. The additional responsibilities and duties that may arise as a consequence could devolve into a new, competent and empowered supervisory organisation (e.g. WSO) rather than the existing small set-up of the OOSA for limited tasks.

Growth of Space Debris

The first recorded man-made object to intrude into outer space was *Sputnik-I* of the former USSR in 1957. That very moment, the pristine nature of outer space environment had been technically disturbed. Thereafter, many more Soviet Sputniks and US launch rockets and artificial satellites forayed into the outer space; and now there are many more spacefaring countries, including India. Some of the satellites launched are still functional and are orbiting the earth and elsewhere undertaking scientific exploration and providing utilities to mankind on the Earth. Many of the launched vehicles and dysfunctional satellites are orbiting in outer space and causing further spread of debris due to explosion or implosion.

An estimate of space debris will be helpful for our understanding of its seriousness. As per the UN registry, 5,196 launches have been made during the period 1957-2011, of these 4,769 have been successful that have carried 7120 payloads.[34] Further, the current rate of launchings per year, since 2011, is about 80[35] and after due extrapolation, as of now a total of nearly 10,000 launches have taken place with almost 8,000 satellites placed in outer space belonging to almost 100 countries. Of these, nearly 1,500 satellites are presently operational in outer space. It can thus be easily deduced that granting all possibilities, nearly 6,000 satellites are non-functional and as good as space debris. The exactness of the estimates may have been controverted, but that would not shrink the magnitude of the problem.

Undoubtedly, outer space has since become heavily cluttered with detritus particularly in the low-earth orbit, and with dead satellites in the geo-synchronous orbit. The worst factor is that the debris has a long life of thousands or millions of years in space depending upon the altitude of its orbit with little leeway to automatically reduce this. Thus, the debris poses a serious risk to space traffic and narrows the window of opportunity for exploration, peaceful uses and space-based utilities to which our present generations have got addicted to. The detrimental effects of debris are multi-dimensional and critical.

With litter in the outer space increasing by the day, concerns relating to sustainability of its environment are obvious. High density of debris is proving ultra-hazardous to space traffic for peaceful activities and scientific explorations. The risks of collision associated with launches have escalated exponentially and soon a stage may be reached when junk in outer space may touch the unacceptable critical limit. Moreover, adverse repercussions of all this accumulation on the natural ecology of outer space are not yet fully revealed to science, and continuing deterioration may result in a Kessler Syndrome event, and in a way boomerangback.[36] Thus, the survival of humanity is at stake.

This development was not visualised by space scientists and framers of the Treaty either in its qualitative nature or in quantitative terms. The growing pollution of outer space is an unanticipated eventuality and has now come to face us as a stark reality. The OST has no substantive provisions to avoid or reduce future debris or clean up the mess already created. Article IX of the Treaty is an advisory "to avoid harmful contamination, and also adverse changes in the environment of the Earth resulting from the introduction of extra-terrestrial matter and, where necessary, shall adopt appropriate measures for this purpose".

The existing provision hardly appreciates the current magnitude of the problem, leave aside imposing remedial measures. The Mitigation Guidelines were adopted in 2007 as non-binding norms and take care of future inductions. The past congestion still remains unattended though space scavenging technologies are under development in Switzerland and Japan and may be operational in five years. However, the basic need of the time is self-infused space discipline in space activities to assure sustainability of space environment for the future generations.

Suggested Solutions

The OST is 50years old, and over this period, it has acquitted itself very well. It has met the incumbent challenges and has matured enough as a *Grundnorm* of space law, ushering in new jurisprudence. Today, we are at a turning point where new challenges confront us that cannot be effectively tackled within the existing legal framework. In brief, the challenges emerge from the original weaknesses of the Treaty that have existed since inception but are getting magnified under pressure. The other is the burgeoning private enterprise that is literally gate-crashing into the domain of space activities for its vast potential and high profitability.

Finally, the challenge stems from our own careless creation in the form of space clutter and detritus that is detrimental to the sustainability of space environment and narrows the aperture of opportunity to space operations. And the problem is escalating exponentially for several reasons. The existing scenario clamours for solutions, and the legal inadequacies of law appear prominent and serious that cannot be remedied by placebos or interpretational jugglery. It needs a solution, sound and proper. A few viable options are briefly discussed in succeeding paragraphs.

Continuance of Status Quo

In management studies, 'No Action' is also an option and so called 'action or activity' as long as it is a deliberated decision. It perpetuates the existing state of affairs and lets the situation drift unguided on its own force and on its own course under the pressure of circumstances. It is in fact a vote for the status quo, desiring no deviatory action or course correction. It is imaginable to recount the OST as a fount of human creativity or as a superb treaty with novel tenets and durable concepts, but let's also not feel unnecessarily nostalgic or possessive about it. For the OST, however, such a proposal of status quo is not desirable.

On the other hand, there is need to recognise that it is a compelling time for change and we need to read the writing on the wall. The alternative under consideration does not constitute a wake-up call for timely action but inducement to evade stirring issues. Such conscious neglect to eschew action can be detrimental and can only be ignored at our own peril. The situation contains a signal to chaos which may ultimately effects the very survival of humanity. Therefore, this option is negative in character and is not recommended at all. Graffiti says that running away from any problem only increases the distance from the solution. The decision lies with us.

Contemporise the Treaty

On the downside of the Treaty, one can highlight its present demerits of being a vintage law and its unsuitability to adapt itself to the newness of burgeoning space activities of the future. Based on these considerations, one would expect an enlightened New Treaty with futuristic nuances. This may not be practically possible in the near future. Moreover, some of the changes may appear too radical for immediate or easy acceptance.

Therefore, the next obvious solution is amendment of the Treaty, as necessary to upgrade and contemporise it. This involves pursuit of an equally difficult transformational agenda yet a feasible one. This will involve revision of OST to rectify its intrinsic faults and suitably accommodate the compulsions of present scenario with a futuristic stance. It would have to proactively envisage and accommodate anticipable future developments, particularly in respect of technological advances, new commercial avenues, formats of business ventures and the possibility of scavenging of space debris from public highways of outer space. Thus, we need to ferret out acceptable reforms and harmonise such provisions to retain its salience and sustain long-term relevance of the Contemporised Treaty.

In case of acceptance of this option, perhaps the Committee on the Peaceful Uses of Outer Space (COPUOS) may have to undertake the project of revision of the Treaty or constitute a sub-committee for the purpose. Diplomats are well aware that such a process of negotiations, e.g. the revision of a treaty, is laborious, time-consuming and tortuous, particularly in seeking consensus. On the other hand, this process will require patience, perseverance and a spirit of conciliation which may be difficult to muster at this point of time. The required diplomatic drill will be indeed arduous. Therefore, this option does not appear the best solution under the circumstances.

Conclude Specialised Protocols to the Treaty

A viable and attractive proposal emerges as a compromise of earlier options in order to avoid their bottlenecks and handicaps and yet retain their revisionary merit. This option lends an advantage to harmonise the OST by concluding specialised protocols to cater to major areas of revision, e.g. rules relating to mining/harvesting of celestial bodies, legality of private enterprise undertaking space activities and consequences of liability, regulation of space travel and tourism, vision of human migration to celestial colonies, management of space debris through mitigation as well as remediation measures. This solution seems suitable because the prime philosophy of the Treaty remains the same and it needs no intrinsic conceptual changes while only some structural reforms appear necessary to cater to emerging eventualities.

Apart from these, some minor amendments can be made in the Treaty itself to upgrade and make it responsive to the future. Such changes can be introduced in its articles relating to Astronauts as Envoys of Mankind and Prohibition of Bombs Orbiting the Earth. If amicably possible, definitions of important terms could also be introduced in the Treaty. This may appear a patchwork argument, yet seems ostensibly tenable and legally expedient. Monitoring regulations and supervising organisation, as necessary and germane, can be evolved suitably as expedient or with experience.

It is humbly suggested that the UN may consider, as has been its practice in the past, an initiative through the COPUOS to draft suitable protocols to the OST and other appropriate amendments. If deemed fit, COPUOS may form topical sub-committees for this task. Specialised assistance in drafting may also be elicited from renowned domain scholars with ideational commitment and affiliations with renowned institutions like International Institute of Space Law and McGill University or space policy centres like George Washington University and Astropolitics Consulting.

Protocols, it is believed, will be easy to negotiate and formalise and can be separately adopted as when individually ready. Thus, the new legal matrix would be able to progressively accommodate changes and enunciate the second generation of space law that will be perceptibly liberal, intrinsically interactive, responsive to unanticipated contingencies and conducive to latent opportunities that can collectively benefit humanity and enhance global peace and prosperity.

Importance of National Space Legislation

The OST, howsoever adaptive to private enterprise and commercial space activities, shall always acknowledge only sovereign state entities as parties in its core. This aspect comes in focus particularly in areas of state responsibility as well as international liability, irrespective of the juridical identity of the activity performing unit, whether state, public-private partnership or private natural citizen. The OST (Article VI) recognises only national activities as a genre and performers have no independent *locus standi*. Their legitimacy is through the state of allegiance.

This mandate has been further amplified in the OST by asserting, "The activities...shall require authorisation and continuing supervision by the appropriate State Party to the Treaty."[37] This constitutes an overall responsibility of the state for control and superintendence of all activities in outer space and on the Moon including celestial bodies. Such a responsibility would definitely require regulatory controls and verification of compliances through continuous monitoring and this can be best achieved by a suitable National Space Legislation and its effective enforcement. Space law needs a quantum jump at national level.

As on today, over 20 countries already have enacted such laws, some of these are sketchy and brief while some are fairly detailed and elaborate like the ones by the US and Australia. India has not yet introduced such a bill in the Parliament though the Indian Space Research Organisation (ISRO has taken up the initiative of drafting the same. There are many others also who are yet to undertake this legislative exercise. Nevertheless, the efficacy of domestic space law in tandem with the Treaty can hardly be underrated while the need for such action and on priority can hardly be over-emphasised. In fact, it seems imperative in view of the growing complexity of space operations. Divided responsibility can be expected to be better discharged.

Convert OOSA into a UN Organ

Whichever of the above solutions is found suitable and acceptable, one point becomes clear and constant that the future changes to the Treaty are going to generate a lot of extra and specialised workload of governance of outer space which OOSA may not find manageable nor competent to handle professionally. Frankly, even in the present set-up, Treaty execution and enforcement is a weak link. Therefore, the existing OOSA is unequal to the future responsibilities and additional tasks due to be allotted in consequence of expanding commercial space activities, increasing participation of private enterprise and continued generation of space debris.

The additional generic responsibilities would relate to regulatory duties, inspection of compliances, space scavenging and quasi-judicial pronouncements in dispute redressal. Examples can be enforcement of the Treaty, grant of mining leases on the asteroids and the Moon and other celestial bodies, possible levy of a financial cess on launches to cover the costs of cleaning up of the space environment and other incidental tasks like sharing of benefits. Further, the provisions of Article XIII of the OST are not adequate enough for redressal of disputes either by procedural mechanism or institutional set-up. Therefore, the proposed organisation will have to act as a forum of conciliation and redressal of first instance.

There is thus an undoubted need to address these issues appropriately so that the law is not misinterpreted by states or overtaken by technology or public zeal for space travel and resultant contingencies. Therefore, law sorority must also cognise the ensuing developments and be pro-active in legal thought and responsive to treaty initiatives. This is a clarion call and the time is now to cogitate on impending challenges, address them squarely, find judicious solutions and act decisively. Let us not be sieged by inaction in the face of impending chaos.

In a way, the situation, arising out of a few arbitrary and abberative initiatives by some states[38] and occurrence of space collisions[39] seemed pregnant with conflict. Such repeats would needs a vigilant, empowered and potent set up in the form WSO, which can be treated as Regulator, Trustee and Inspector of outer space and celestial bodies. Its functions and roles may encompass broad-spectrum duties of space governance, competent inspection of compliances, sustenance of space environment and quasi-judicial dispute redressal forum. Within the structure of WSO, there could be an Information Exchange Bureau (IEB)[40] to disseminate information shared by states after exploration. There is need to build up political lobby, diplomatic thrust and legal opinion towards formation of WSO; and at the same time initiate planning of its complexion and structure as well as commence preparation towards its establishment by absorbing the OOSA.

Conclusion

Times have changed, technology has advanced, private enterprise has come of age, harvesting of natural resources from celestial bodies has become feasible, space tourism is just round the corner and states are becoming more reticent in space investments and attitudinally more aggressive, while space law has remained nearly static over last half a century. The OST, therefore, needs an overhaul to again become relevant to space governance. For the purpose, a recommended proposal is to draft and adopt *specialis* protocols to effectively handle each development and issue separately and bring the respective protocol in force as and when ready. This possibly would be the fastest way out of the present legal morass and towards the second generation of the OST.

For proper re-priming of the OST, the COPUOS needs to seriously work on updating itself through specialised protocols to become contemporary and resilient, as also futuristic and pro-active. This option has been suggested because introducing amendments to the OST will be a difficult, laborious and long-drawn diplomatic procedure while the negotiation of Specialised Protocols is a short-cut and will be relatively easy and fast and each protocol can be independently adopted as and when ready after due processing. The COPUOS may consider forming subcommittees of multidiscipline specialists to look at specific issues threadbare and from all angles and from futuristic perspective.

In conclusion, one recalls Plato's wisdom, his words that we can easily forgive a child who is afraid of the dark; the real tragedy of life is where men are afraid of the light. We are in a similar situation where with the experience of half a century of OST's existence and a fairly clear vision of new technologies and overtures by private business in areas not so explicitly authorised, we are knowingly defaulting on action to update and revise the OST to eschew conflict on the earth and infuse harmony in space activities. One wonders if we are intentionally making the OST fail in its objective, get poor in spirit and fall in its prestigious standing, or it is sheer inertia in action or lack of will power to take the first step. Let's not forget that even a journey of thousands of miles starts with a first step.

ENDNOTES

1. Treaty on Principles Governing the Activities of States in the Exploration and Use of Outer Space, including the Moon and other Celestial Bodies, 1967. In short, the Outer Space Treaty (OST).
2. Ibid., Article II.
3. See http://www.govtrack.us/congress/bills/113/hr 2617/text.
4. OST, No. 1, Article II.
5. For a detailed analysis on the subject, refer Kumar Abhijeet, "Appropriation of Space?—Apollo Lunar Landing Legacy Bill as a Trigger for Colonisation of Space?", *German Journal of Air and Space Law,* 64, 2015.
6. Steven Freeland and Ram Jakhu, "Article 2", in Stephan Hobe et al. (eds.), *Cologne Commentary on Space Law*, Volume 1, *Outer Space Treaty*, Carl Heymanns Verlag, Cologne, 2009, pp. 48 and 54.
7. OST, No. 1, Preamble.
8. Ibid., Article I.
9. Ibid.
10. Ibid., Article I, Paragraph 1.
11. Detlev Wolter, *Common Security in Outer Space and International Law*, UN Institute of Disarmament Research, Geneva, 2005, p. 85.
12. OST, No. 1, Article I.
13. Ibid., Article IX.
14. Ibid., Article I.
15. Ibid., Article IX.
16. OST, No. 1, Article I.
17. Ibid., The Preambular paragraphs.
18. Ibid., Article 1.
19. D. Goedhuis, "An Evaluation of the Leading Principles of the Outer Space Treaty of 27 Jan 67", *NTIR*, 15, 1968, p. 40.
20. Ibid., p. 251. Rüdiger Wolfrum, "The Problems of Limitation and Prohibition of Military

Uses of Outer Space", *ZaoRV*, 44, 1984. Also refer R. Wolfrum, "Common Heritage of Mankind", *Encyclopaedia of Public International Law*, 11, 1989, p. 67.

21. R. Dolzer, "International Cooperation in Outer Space", *ZaöRV*, 45, 1985, p. 527. Also refer note 18 ante, p. 92.
22. US Commercial Space Launch Competitiveness Act, HR 2262. Ostensible object of this Act is to encourage sustained development of space and create great economies of history.
23. G.P. Zhukov, "The Problem of the Definition of Outer Space", , *Proceedings of the Tenth Colloquium on the Law of Outer Space*, 271, 1967, p. 273. Even the scientific community has dithered on definitions and has been changing them on new discoveries.
24. This subject has been discussed in detail in: G. S. Sachdeva, "Astronauts: Envoys of Mankind in Outer Space", *Outer Space: Security and Legal Challenges*, KW Publishers, New Delhi, 2010, pp. 89-102.
25. OST, Article IV.
26. Under FOBS a bomb carrying vehicle does not remain in entire orbit around the Earth but for a part of it and thus does not violate the Treaty provision.
27. OST, Article IV.
28. G.S. Sachdeva, *Outer Space: Security and Legal Challenges*, KW Publishers, New Delhi, 2010, pp. 161-85.
29. Ibid., pp. 89-102.
30. *Dictionary of Technical Terms for Aerospace Use,* Washington DC, 1965.
31. *The Times of India*, July 5, 2016.
32. For example, Woerner's Moon Village.
33. US Commercial Space Launch Competitiveness Act, No. 23.
34. http://claudelofleur.ac.co/spacecraftsindex.html.
35. For data on country-wise and year-wise launches made, see "Orbital Launches of 2013", Gunter's Space Page at http://space.skyrocket.de/doc chr/lau2012.htm (Accessed January 15, 2014). Also refer Manpreet Sethi, "Limitations in the Existing Legal Framework in Outer Space", *Asian Defence Review*, 2013.
36. This critical effect was hypothesised by NASA scientist Donald Kessler. The syndrome visualises a scenario in which orbital trash eventually gets accumulated in outer space to such an extent of concentration and over-population to reach a threshold of criticality where even a single accidental collision or self-explosion can result in a bigger blast setting off a chain reaction of accidents and collisions making it well-nigh impossible to navigate in space. A similar scenario has been depicted in a movie titled *Gravity* (2013).
37. The OST, Article VI.
38. ASAT experiments by China and the US in 2007 and 2008, respectively.
39. The Iridium collision with Soviet spaceship.
40. IEB could work on the lines of International Standards Organisation (ISO) where National Space Organisations and corporate or private space ventures could seek membership.

4

Relevance and Limitations of Outer Space Treaty in 21st Century

Ranjana Kaul

Who owns space is an ancient and existential question. Since the 1967 Outer Space Treaty (OST)[1] set out the universally accepted rules for the game[2], this question has been debated intermittently, albeit with geopolitical dilemmas and commercial opportunities having replaced the earlier philosophical rigour. The 50th anniversary re-assessment of the Treaty framework will most likely centre round the 2015 legislative fait accompli granting private property rights in space resources to its citizens[3]. The immediate response to the US commercial space mining law is fractured by opposite opinions on the choice of rules of interpretation of statute, to analyse the correctness or otherwise of the US domestic law in context to OST international obligations.

The Treaty has sustained the past five decades demonstrating dexterity of the United Nations (UN) Member States to find solutions despite the many threats and challenges. But the Treaty is not a complete code. It does not deal comprehensively with all possible types of space activities which are necessarily driven by "the developmental speed and ever-growing physical reach of space technology."[4] Indeed, the only real challenge to human ability to extend physical reach in outer space are the Laws of Physics. Today, as never before, we are acutely conscious of this limitation. Therefore, it is of utmost importance that the question must be debated in all forums, including the Committee on Peaceful Uses of Outer Space (COPUOS)[5] and the UN

General Assembly (UNGA), whether existing Treaty provisions can be interpreted or amended, to allow for the exercise of private property rights by citizens of member states, for purpose of commercial exploitation, such that the Article II [6] principle remains inviolate and the Article I[7] principle remains un-compromised, directly and indirectly? Arguably, a legislature elected in the future may well have a very different appreciation and sense of the facts and circumstances subsisting at the relevant time.

The debate on *who owns space* and *non-appropriation of outer space* brings to mind the attempt by eight Equatorial states to claim rights over the geostationary orbit[8] by way of the 1976 Declaration of Bogotá, which was rejected by the UN because it ran counter to the OST and did not acquire widespread acceptance[9]. However, in the recent years, there has been a shift away from the recognition of outer space as a common area free of state sovereignty under international law. This has been particularly evident in efforts to '*address goals of space sovereignty*' and to '*establish international space sovereignty policy*' in a '*Space Faring Nations Treaty*' which is intended to guarantee the '*protection of national (commercial) space assets*'.[10] Be that as it may, the abolition of the non-extension of the principle of state sovereignty to outer space could only be possible with the consent of states parties to the OST. Any plan for '*space superiority*' will be contrary to the clause of the OST mandating the use of outer space for the benefit of all mankind, and its obligation to use outer space in the interest of all states.[11]

The US Space Resource Exploration and Utilization Act, 2015,[12] which seemingly establishes extraterritorial reach of the US law by granting ownership rights to private entities in space resources derived from commercial mining on asteroids and other abiotic resources in situ in outer space, triggered instant debate. Reactions were swift and diametrically opposed. Several experts argued in favour of interpretation *lex ferenda*, i.e. an interpretation that supports development of new law,[13] in this case – the US space mining law. Meanwhile, the International Institute of Space Law published a position paper on space resource mining in December 2015.[14]

Other experts consider the US law dangerous and potentially illegal, arguing that it strikes at the roots of the Treaty arrangement – (i) the principle of space, *for the benefit and in the interests of all countries*;[15] and (ii) the principle of *res communis omninum*, i.e. outer space as an area open for free exploration and use by all states which is not subject to national appropriation.[16] According to Ram Jakhu, Director, Institute of Air and Space Law at McGill University, "The 1967 Outer Space Treaty, signed by the U.S., makes it clear that the surfaces and contents of asteroids and other celestial bodies are protected from commercial harvesting; harvesting of space-based resources must be done '*for the benefit and in the interests of all countries*'. In his view the overarching purpose of the treaty leads to the conclusion that '*there really shouldn't be any private property rights in outer space*'."[17] Only time will validate which rule of interpretation of statute, *de lege lata* or *lex ferende,* will yield beneficial results for the global community by ensuring the use of space for peaceful purpose, while yet advancing technological developments and innovations.

Stepping aside from the on-going debate, it would be useful to recall that the international treaties on the law of outer space[18] are the products of the UN mechanism of decision by consensus and duly adopted as such by the UNGA.[19] However, as is self-evident, no new space treaty has followed the 1979 Moon Agreement.[20] The vacuum created by the absence of consensus after 1979, for new binding space law treaties, was filled in by the so-called '*soft law*', i.e. an expression of a common approach agreed to by state parties in the COPUOS related to new technological developments and other consequences incidental thereto in draft resolutions adopted by the UNGA. Soft law is a euphemism commonly used for the non-binding UNGA resolutions aimed at assuring safe, secure and sustainable use of outer space. Some resolutions are adopted as a response to space activity by an individual state that has the potential to harm the safe, secure and sustainable space activities of other state parties. Some representative examples are: (i) *Guidelines related to space debris mitigation, 2007*;[21] (ii) *no first placement of nuclear weapons/weapons of mass destruction (WMD) in outer space, 2014*;[22] and (iii) *international cooperation on peaceful uses of outer space, 2015* [23] which reflects the reality of geopolitical challenges influencing the use of space. In this connection, it is pertinent to note that a 2004 UNGA resolution[24] encourages member states to enact national space law compliant with provisions of the international space law treaties so as to facilitate countries to fulfil international space treaty obligations.

It is easy to see that the subject matter of UN resolutions since 1980 tell the story of how the international community collectively perceived and dealt with existing and emerging threats to the continuous access to, and safety, security and sustainable use of outer space. The problem, of course, is that UN resolutions are non-binding and, despite being adopted by consensus,[25] cannot ensure, or compel compliance by individual member states. The same is true of the international space treaties which do provide verification or monitoring mechanisms, nor do violations of treaty provisions impose adverse consequences on the non-compliant member state.

Indeed, as far as is known, no spacefaring country has sought prior *appropriate international consultation*, required under Article IX[26] provisions, where it had *reason to believe that an activity or experiment planned by it or its nationals in outer space, including the Moon and other celestial bodies, would cause potentially harmful interference with activities of other states parties*[27] even if it was reasonably certain that a proposed 'experiment' would add significant amounts of debris that would cause adverse changes in the space environment. It is therefore obvious that, going ahead, the success of international space treaties will continue to depend primarily on individual actions in support of collective resolution to ensure greater good. It would seem, as if, we have come a full circle, back to the philosophical argument.

In light of the aforesaid, the US Space Resource Mining Act is the first known instance of domestic legislation on space activity, which grants its citizens actionable rights that traverse well beyond the core principles of the OST, albeit, with a disclaimer of extraterritorial sovereignty expressed as *Sense of the Congress* statement therein. Arguably, a plain reading of Article II suggests that it cannot be interpreted except by the *de lege lata* rule, i.e. by stating the *law as it is*. The language of Article II is unambiguous. Outer space is *not subject of national appropriation by claim of sovereignty by means of use or occupation or by any other means.* The threat to the principle of *non-appropriation of outer space* has been written almost since the early years of the Treaty.[28] Unquestionably, the space mining law specifically recognises private property rights in space resources on asteroids and outer space is the outcome the exercise sovereign authority by the US. Could it be discounted, therefore, that in the future lawmakers will not understand or interpret the provisions of the Space Mining Act to recognise it as the basis to claim extraterritorial sovereignty in outer space, notwithstanding the *Sense of the Congress* statement expressed in 2015[29]. In this view of the matter, it would be pertinent to ask in what manner the COPUOS could deal with the likely increase in enactments of national law by ratifying states, which traverse contrary to international space law treaty provisions, simultaneously disclaiming intention to engage in contravention admitted into the substantive law.

The impact of the US space mining law has been immediate. The Duchy of Luxembourg[30] has proposed a comprehensive legislation, to be made effective in 2017, which unlike the US law will be applicable to its nationals and also to international corporations. The law will issue space resources-dedicated licences, and government supervision of the activities of operators and regulating their rights and obligations will be ensured by Luxembourg in accordance with the OST. Already, two US companies, Deep Space Industries Inc. and Planetary Resources Inc. have established themselves as "legal entities" in Luxembourg. While Deep Space is working on designing launch vehicles for asteroid mining, Planetary Resources, a start-up owned by Google co-founder Larry Page, plans to develop satellites for exploration.

It is clear that the US has taken the first step to expand opportunities for space commerce to benefit US enterprises, beyond the boundaries prescribed by the OST. And, Luxembourg proposes to leverage its tax friendly laws to attract international space corporations to establish legal entities in its jurisdiction. It is undeniable that the lucrative commercial potential of outer space waits to be exploited by any country that command technological and financial capability.[31] In 2014 the global space industry was estimated worth US$ 330bn. Commercial mining of asteroids offers a new and lucrative vista. Is space resource mining the new high table in the geopolitics of outer space? In these circumstances, is the post facto interpretation *lex ferenda* justified, if only in recognition of ground reality of realpolitik? Alternatively, will the *lex lata* interpretation actually prove detrimental to new and future developments in space science and space technology? Would the *lex lata* interpretation impede new developments in space commerce? Which interpretation is in the best interest of the space environment and the Earth? Do such new national laws contain seeds that will negatively impact on global space security in the 21st century?

For sure, the future of international space treaty provisions in the 21st century shall be subjected to intense geopolitical pressure, as it was during the drafting of the treaty culminating in 1967. The only difference is that as against two space powers in 1957, in 2017, on the 50th anniversary of the 1967 OST, there are 11 space powers with indigenous space launch capability. Of these five are major space powers. Over 60 countries access space. Furthermore, several wealthy non-space powers are deploying their considerable financial resources to set up national space programmes. And, almost every country has space capability as an essential feature of its national security architecture. It is without doubt the triumph of the OST and international community.

But, as space gets crowded and global conflicts on the Earth multiply manifold in complexity and dimensions, challenges to the existing international space law regime will be linked intrinsically to the well-known conflict of interest between national objectives and international obligations arising qua space security. The challenges, including overcrowded orbits; an ever-growing demand for scarce radio spectrum, especially for military use and cybersecurity will be juxtaposed against the universally recognised imperative for international cooperation to achieve common objectives like addressing climate change and disaster management and fighting international terrorism dressed in different types of sinister cloaks.

The OST sets rules for the peaceful use of outer space, where '*peaceful*' is understood to mean '*non aggressive*'. Article VI,[32] the non-armament clause, prohibits member states from placing, installing and stationing nuclear and WMD in the orbit of Earth, on the Moon and other celestial bodies. As such, although space technologies were developed for military purpose, the Treaty specifically deals with its application for civil purpose through the mechanism of the COPUOS and UNGA. Admittedly, the dual use ability of space technologies is the proverbial double-edged sword. Space '*armament*' aspects are subject matter in the Conference on Disarmament (CD) specifically in the context of prevention of arms race in outer space (PAROS) and prevention of weaponisation of outer space.

Indeed, space has always been a domain restricted for military use, and military space technologies were enlarged for civilian use only after 1990-91. Although, the US and USSR did not breach Treaty provisions even after 2002 when US withdrew from the 1972 Anti-Ballistic Missile Treaty,[33] albeit it allowed the US freedom to develop missile defence including space-based systems. Additionally, the US has resolutely opposed a Treaty on PAROS in the CD. Then President George W. Bush articulated the US approach to outer space in the 2006 US National Space Policy. The policy rejected arms control agreements that might limit US flexibility in space and asserted a right to deny access to space to anyone "hostile to US interest",[34] arousing global concern about the possibility of application of the doctrine of pre-emption in outer space.

On January 11, 2007, China successfully carried out an anti-satellite (ASAT) weapon test,[35] becoming the third country after the US and Russia to have demonstrated ASAT kill capability. The immediate adverse impact was that the ASAT test introduced a worrisome debris cloud of several thousand debris pieces and particles into space environment that will remain in the Earth orbit for several centuries.[36] China said that it did not know that so much debris would be introduced into the fragile space environment.

Importantly, the Chinese ASAT test had a cathartic impact, in the context of the other significant events that followed. First, there was negative international reaction to the ASAT test which was a demonstration of Chinese technological capability to intercept moving objects in outer space. It did not matter that China said the test was not aimed against any particular country.

Second, the UNGA endorsed the *Space Debris Mitigation Guidelines* in December 2007.

Third, the US which had not conducted any ASAT tests since 1985, conducted *Operation Burnt Frost*, the code name given to the military operation to intercept and destroy a non-functioning US National Reconnaissance Office (NRO) satellite named USA-193[37] on February 20, 2008. That test also introduced many pieces of debris into space environment. The US said that they had destroyed their own satellite to save the space environment from further contamination by its toxic fuel which was leaking from the satellite fuel tank.

Fourth, the CD witnessed a multilateral effort in February 2008 when Russia and China tabled for consideration of members the draft *Treaty on Prevention of the Placement of Weapons in Outer Space and of the Threat or Use of Force against Outer Space Objects* (PPWT), a legally binding treaty that would outlaw the weaponisation of space and rooted in the Article IV provisions. The US opposes the PPWT.

Fifth, in 2008, again, the EU sought support for a binding Code of Conduct for Space Activities, which aimed, among other things, to reduce or eliminate creation of space debris and foster space travel. In 2010, the EU Code was transformed into the International Code of Conduct for Space Activities which lists best practices for states to follow on voluntary basis.

In this background, a failure to achieve unanimous vote in favour of the December 7, 2015 UNGA resolution on '*no first placement of weapons in outer space*',[38] which reiterates Article IV provisions is not surprising. Although 129 countries voted in favour of the resolution, the EU chose to abstain, while the US, Israel, Georgia and Ukraine voted against it. The voting on the resolution reflects well-known positions in the CD. It also reflects the well-known approach to unchallenged freedom of action in outer space that is now reflected in several national space policies or national security policies across the world.

The juxtapose between accelerated development of space weapons technology backed by national security policies in which space capability is an essential pillar; together with the prospect to private commercial and industrial, entrepreneurship in outer space, including the Moon and other celestial bodies, have potential to become the most difficult challenges in the 21st century. The OST has all the elements that allow for the peaceful use of outer space by all countries on non-discriminatory basis. Yet, it is difficult to ignore lessons from history which tell of how armies marched to protect commercial enterprise established by their citizens and to establish control over territories that yielded natural resources, far away from their home countries.

Amidst this rather ominous and seemingly doomsday scenario is the undisputed fact that outer space technologies provides nations with the best tools to combat challenges of poverty, socio-economic development, employment generation, education, medicine and empowerment of citizens. The UN has succeeded in enabling international cooperation to harness technology and identify internationally acceptable solutions to help countries manage the complex array of adverse impacts in every aspect of their national life, particularly mechanisms for managing climate change for long term sustainable development and for disaster management caused by increasing numbers of natural calamities. Indeed the UN Security Council recognised in April 2007 that "the threat that the aggregate impacts of climate change might cause, with not only serious environmental, social and economic consequences, but also implications for peace and security".[39] One of the most contentious challenges has been the sharing of geographical information derived from satellite images, i.e. remote sensing (RS) data. Although the 1986 *UN Principles Relating to Remote Sensing of the Earth from Space*[40] require countries participating in remote sensing activities to carry out activities in accordance with the OST and in terms stated particularly in Principles X; XII; and XII. However, the Principles do not contemplate or refer obligation to provide RS data for climate change management and disaster management. Indeed, RS data relevant for climate change, sustainable development and disaster management are new and complex legal concepts in international space law and for national law because of its obvious national security implications. Going forward, therefore, it is important for the UN to redouble efforts to arrive consensus on definitions of key terms to lend clarity and legal certainly and also to ensure accessibility and affordability to reliable, updated RS data relevant for climate change management, disaster management and sustainable development.

The future of the OST in the 21st century depends on the countries who must work it ceaselessly, individually and collectively. The Treaty provisions have the inherent flexibility to help formulate mechanisms and solutions appropriate to deal with competing demands of contemporary times. The Vienna Convention on the Law of Treaties[41] states that international treaties and obligations must be fulfilled in good faith: *pacta sunt servanda*, and that internal law cannot be used as justification for failure to perform the treaty.

ENDNOTES

1. Treaty on Principles Governing the Activities of States in the Exploration and Use of Outer Space, including the Moon and Other Celestial Bodies [hereinafter referred to as the OST or Treaty], 1967; UNGA doc A/6431, at http://www.unoosa.org.
2. Ibid. See Preamble:
   Desiring to contribute to broad international cooperation in the scientific as well as the legal aspects of the exploration and use of outer space for peaceful purposes.
   And Article XI:
   In order to promote international cooperation in the peaceful exploration and use of outer space, States Parties to the Treaty conducting activities in outer space, including the Moon and other celestial bodies, agree to inform the Secretary-General of the United Nations as well as the public and the international scientific community, to the greatest extent feasible and practicable, of the nature, conduct, locations and results of such activities. On receiving the said information, the Secretary-General of the United Nations should be prepared to disseminate it immediately and effectively.
3. Space Resource Exploration and Utilization Act of 2015, Title IV of the US Commercial Space Launch Competitiveness Act of 2015 (H.R.2262), passed by US Congress and signed into law by President Barack Obama on November 25, 2015, at https://www.congress.gov/bill/114th-congress/house-bill/2262/text. See Asteroid Resource and Space Resource Rights:
   A United States citizen engaged in commercial recovery of an asteroid resource or a space resource under this chapter shall be entitled to any asteroid resource or space resource obtained, including to possess, own, transport, use, and sell the asteroid resource or space resource obtained in accordance with applicable law, including the international obligations of the United States. [Referred to as 'US Space Mining Act']
4. A.G. Koroma, "The Development of International Law and the Peaceful Uses for Outer Space", Third Nandasiri Jasentuliyana Keynote Lecture on Space Law, International Institute of Space Law (IISL), 2011, at http://www.iislweb.org/docs/2011keynote.pdf.
5. A permanent body of the UN General Assembly (UNGA).
6. OST, No. 1. See Article II:
   Outer space, including the moon and other celestial bodies, is not subject to national appropriation by claim of sovereignty, by means of use or occupation, or by any other means.
7. Ibid. See Article I:
   The exploration and use of outer space, including the Moon and other celestial bodies, shall be carried out for the benefit and in the interests of all countries, irrespective of their degree of economic or scientific development, and shall be the province of all mankind.
   Outer space, including the Moon and other celestial bodies, shall be free for exploration and use by all States without discrimination of any kind, on a basis of equality and in accordance with international law, and there shall be free access to all areas of celestial bodies.
   There shall be freedom of scientific investigation in outer space, including the Moon and

other celestial bodies, and States shall facilitate and encourage international cooperation in such investigation.

8. Declaration of the First Meeting of Equatorial Countries [also known as the "Bogotá Declaration"], adopted December 3, 1976, at http://www.jaxa.jp/library/space_law/chapter_2/2-2-1-2_e.html. The representatives of the states "traversed by the Equator" met in Bogotá, Colombia, to discuss their rights over a natural resource which – to them – had been unfairly removed from their sovereignty. Specifically, these states felt that their rights to control natural resources had been unfairly abridged by the OST, which cements the rule that "outer space, including the Moon and other celestial bodies, is not subject to national appropriation".

   At the conclusion of the conference, Colombia, Ecuador, Congo, Indonesia, Kenya, Uganda, and Zaire, with Brazil as an observer, proclaimed the Bogotá Declaration to assert their rights. To get around the OST's declaration that outer space is not subject to national appropriation, the Bogotá Declaration categorised the geostationary orbit (GEO) as a natural resource, not a region of space. Indeed, the Declaration claimed that the unique properties of GEO are created by the Earth itself. By categorising this orbit as a resource, these states could call on the *jus cogens* principle that states have absolute control over their natural resources. Ultimately, however, the Bogotá Declaration was precluded by public international law. Article II of the OST has passed into customary international law and the free use and non-appropriation of outer space, through over 60 years of state practice, is enshrined in public international law. Consequently, adopting the Declaration's view on GEO would contravene well established limits on state sovereignty.

9. Celestial real estate on sale: In contrast to the Bogotá Declaration which was an effort by eight Equatorial countries to claim the GSO, there are instances since 1980 when private US citizens started selling extraterrestrial real estate. An early example of private individuals claiming celestial property or selling it to others first came to public knowledge in 1980 with Dennis Hope, an American entrepreneur who claimed that he owned the Moon and began selling extraterrestrial real estate for around US$ 20 per acre. Also the Lunar Land Company which advertised itself as *the world's most recognised celestial real estate agency* claimed the legal right to sell land on the Moon based on the myth that the OST prohibited only governments from owing extraterrestrial property, as demonstrated by the fact that OST fails to mention individuals and corporations. Needless to say, the entrepreneurs failed to inform clients that the OST imposes international responsibility on member states to authorise and continuously supervise its own activities and also activities undertaken by its non-government entities, i.e. individuals and corporations.

   See Rachel Hardwick, "Dennis M. Hope Has Owned the Moon since 1980 because He Says So", *Vice*, April 11, 2013, at http://www.vice.com/read/ive-owned-the-moon-since-1980 and Lunar Land Company, at https://www.lunarland.com/.

10. US "Long –Range Plan", *Implementing USSPACECOM Vision 2020*, at http://www.fas.org/spp/military/docops/usspac/lrp/toc.htm.

11. A. G. Koroma, No. 4.

12. Space Resource Exploration and Utilization Act of 2015, Title IV of the US Commercial Space Launch Competitiveness Act of 2015, No. 3.

13. When interpreting a statute, the principle of legality is foundational in context to customary international law. The OST is an international treaty that primarily derives from customary international law and also the Charter of the United Nations. The OST establishes principles based on customary law and validated by 50 years of state practice to keep access to, from and in outer space, including the Moon and other celestial bodies is open for use for peaceful

purpose on the basis of non-discrimination. However, the focal problem for interpretation relevant to *lex lata* versus *lex ferenda* approaches is that the borderlines between interpretation of existing law and the making of new law are inevitably fluid.

14. "Position Paper on Space Resource Mining", IISL, December 2015, at http://www.iislweb.org/docs/SpaceResourceMining.pdf.
15. OST, No. 1, Article I.
16. The non-appropriative nature of outer space, first declared in the UN General Assembly Resolution 1721 and 1962, was formally laid down in Article II of the 1967 OST. For detailed commentary, see Fabio Tronchetti, "The Non-Appropriation Principle under Attack: Using Article II of the Outer Space Treaty in Its Defence", IISL, 2007, at http://www.iislweb.org/docs/Diederiks2007.pdf.
17. R. Ridderhof, "Space Mining and (US) Space Law", *Library Blog*, Peace Palace Library, December 18, 2015. Also see Philip Ball, "Time to Rethink the Outer Space Treaty", *Nature*, October 4, 2007, at http://www.nature.com/news/2007/071004/full/news.2007.142.html.
18. There are five international space treaties that together make up the corpus of public international space law. The 1967 OST, the umbrella framework that established the fundamental principles that govern the conduct of human activities in outer space, was followed by four other treaties each expanding on various important principles. These are (i) Rescue Agreement, 1968; (ii) Liability Convention, 1972; (iii) Registration Convention, 1975; and (iv) Moon Agreement, 1979.
19. Article 10 of the UN Charter defines the Functions and Powers of the General Assembly, "The General Assembly may discuss any questions or any matters within the scope of the present Charter or relating to the powers and functions of any organs provided for in the present Charter, and,...may make recommendations to the Members of the United Nations or the Security Council or to both on any such questions or matters." In other words, resolutions adopted by the General Assembly on agenda items are considered to be recommendations and are not legally binding on the Member States. The only resolutions that have the potential to be legally binding are those that are adopted by the Security Council.

    This explains why Member States consider it so important to adopt a resolution that has the widest possible agreement among them. Before taking action on a draft resolution, they spend hours discussing every word in the resolution in the hope of reaching agreement on the text. When consensus on the text is reached all of the Member States agree to adopt the draft resolution without taking a vote. Adopting a draft without a vote is the most basic definition of what consensus means. If 192 Member States agreed on the text but there is just one Member State that requests a vote, then consensus is not reached.

    If a General Assembly resolution is not legally binding, then the best way to encourage all Member States to implement the recommendations expressed in a resolution is to get all of them to agree on the same text. When a resolution is adopted by a simple majority, those that did not vote in favour of a resolution on a particular agenda item will be less likely to implement the actions on an agenda item that are recommended in a resolution.
20. Agreement Governing the Activities of States on the Moon and Other Celestial Bodies [the Moon Agreement], 1979, at http://www.unoosa.org/oosa/en/SpaceLaw/gares/html/gares_34_0068.html. The 1979 Agreement applies to the Moon and all other celestial bodies within the solar system other than the Earth, including orbits or other trajectories to or around them. It turns over jurisdiction to the international community, declares that the Moon is subject to non-appropriation, non-weaponisation. Its natural resources are the common heritage of mankind, as such cannot become the property of any state, and that

states will establish an international regime, including appropriate procedures, to govern the exploitation of the natural resources of the moon as such exploitation is about to become feasible. The Moon Agreement entered into force in 1984, but remains the least ratified of the five space treaties. Excepting France, no other major space power has ratified the Moon Agreement, 1979.

21. UNGA Resolution A/RES/62/217, December 22, 2007, was adopted after the January 11, 2007 ASAT test by China.
22. UNGA Resolution A/69/438, December 7, 2014, was a Russian initiative together with 40 other countries. In all, 129 countries voted for the Resolution while the European Union (EU) abstained from voting; the US, Israel, Georgia and Ukraine voted against the resolution. The 2014 UNGA Resolution appears to have created a link between the non-weapon/ non armament clause, i.e. Article IV of the OST, which is subject matter of non-aggressive and peaceful use of outer space, and the paralysed discourse on military use of space in context to the PPWT and the prevention of arms race in outer space (PAROS) in the Conference on Disarmament where the US, Israel, Georgia and Ukraine have consistently voted against a Treaty on PAROS.
    Also see OST, No. 1, Article IV.
23. UNGA Resolution A/70/495, December 9, 2015, at http://www.un.org/en/ga/70/resolutions.shtml
24. UNGA Resolution A/RES/59/115, December 10, 2004, at http://research.un.org/en/docs/ga/quick/regular/59.
25. UNGA Resolution A/62/217, December 22, 2007.
26. OST, No. 1. See Article IX:
    In the exploration and use of outer space, including the Moon and other celestial bodies, States Parties to the Treaty shall be guided by the principle of cooperation and mutual assistance and shall conduct all their activities in outer space, including the Moon and other celestial bodies, with due regard to the corresponding interests of all other States Parties to the Treaty. States Parties to the Treaty shall pursue studies of outer space, including the Moon and other celestial bodies, and conduct exploration of them so as to avoid their harmful contamination and also adverse changes in the environment of the Earth resulting from the introduction of extraterrestrial matter and, where necessary, shall adopt appropriate measures for this purpose.
    If a State Party to the Treaty has reason to believe that an activity or experiment planned by it or its nationals in outer space, including the Moon and other celestial bodies, would cause potentially harmful interference with activities of other States Parties in the peaceful exploration and use of outer space, including the Moon and other celestial bodies, it shall undertake appropriate international consultations before proceeding with any such activity or experiment. A State Party to the Treaty which has reason to believe that an activity or experiment planned by another State Party in outer space, including the Moon and other celestial bodies, would cause potentially harmful interference with activities in the peaceful exploration and use of outer space, including the Moon and other celestial bodies, may request consultation concerning the activity or experiment.
27. Ibid.
28. Fabio Tronchetti, No. 16.
29. Space Resource Exploration and Utilization Act of 2015, Title IV of the US Commercial Space Launch Competitiveness Act of 2015, No. 3.
30. The Duchy of Luxembourg is drawing up a comprehensive legislation to provide a legal framework that will spur investment in exploiting resources in Near Earth Objects (NEOs)

– the scientific term for asteroids as well as comets. The law is expected to take effect in 2017. According to the Ministry, the law will guarantee operators the right to resources harvested in outer space in accordance with international law. Space resources-dedicated licences will be issued under the new law, and government supervision of the activities of operators and regulating their rights and obligations will be ensured by Luxembourg in accordance with the OST. The Luxemburg law will cover both domestic and international companies. See "Luxembourg Takes First Steps to 'Asteroid Mining' Law", Phys.org, June 3, 2016, at http://phys.org/news/2016-06-luxembourg-asteroid-law.html.

31. The UAE has invested US$ 5.4 bn building space infrastructure and in space technology to build indigenous space capability and the proposed unmanned Mars Mission 2021. The Emirates Institution for Advanced Science & Technology (EISAT) is developing space capability in collaboration with Russia and South Korea. The UAE has sought India's assistance for its proposed Mars Mission.

    Meanwhile, Abu Dhabi has embarked on a project to capitalise on the commercial potential of space tourism. To that end, Abu Dhabi is writing a national space law and building a space port at Al Khaimah.

32. OST, No. 1. See Article IV:

    States Parties to the Treaty undertake not to place in orbit around the Earth any objects carrying nuclear weapons or any other kinds of weapons of mass destruction, install such weapons on celestial bodies, or station such weapons in outer space in any other manner. The Moon and other celestial bodies shall be used by all States Parties to the Treaty exclusively for peaceful purposes. The establishment of military bases, installations and fortifications, the testing of any type of weapons and the conduct of military manoeuvres on celestial bodies shall be forbidden.

    The use of military personnel for scientific research or for any other peaceful purposes shall not be prohibited. The use of any equipment or facility necessary for peaceful exploration of the Moon and other celestial bodies shall also not be prohibited.

33. "The Anti-Ballistic Missile (ABM) Treaty at a Glance", Arms Control Association, January 1, 2003, at https://www.armscontrol.org/factsheets/abmtreaty
34. US National Space Policy, 2006, at https://fas.org/irp/offdocs/nspd/space.pdf.
35. China conducted an anti-satellite missile test on January 11, 2007 in which a Chinese weather satellite, the FY-1Cpolar orbit satellite of the Fengyun series, was destroyed in an head-on-engagement by a kinetic kill vehicle travelling with a speed of 8 km/s in the opposite direction at an altitude of 865 kilometres.
36. T.S. Kelso, "Analysis of the 2007 Chinese ASAT Test and the Impact of Its Debris on the Space Environment", AMOS Conference, Maui, Hawaii, 2007, at https://www.celestrak.com/publications/AMOS/2007/AMOS-2007.pdf.
37. "Navy Missile Hits Dying Spy Satellite, Says Pentagon", *CNN*, February 21, 2008, at www.cnn.com/2008/TECH/space/02/20/satellite.shootdown/. Also see missilethreat.com/china-u-s-conduct-Missile-Defense-Tests.
38. UNGA Resolution A/RES/70/27, December 7, 2015.
39. "UN Security Council Holds First-Ever Debate on Impact of Climate Changeon Peace, Security, Hearing over 50 Speakers", UN Security Council Press Release, April 17, 2007, at www.un.org/press/en/2007/sc9000.doc.htm.
40. UNGA/RES/41/65, December 3, 1986. See Principle X:

    Remote sensing shall promote the protection of the Earth's natural environment. To this end, **States** participating in remote sensing activities that have **identified information in their possession** that can be used to avert any phenomenon harmful to the Earth's natural environment **shall disclose such information to States concerned.** (Emphasis added)

And Principle XI:
Remote sensing shall promote the protection of mankind from natural disasters. To this end, **States** participating in remote sensing activities that have **identified processed data and analysed information in their possession** that may be **useful to States affected by natural disasters,** or **likely to be affected by impending natural disasters,** shall **transmit such data and information** to States concerned as **promptly as possible.** (Emphasis added)

41. Vienna Convention on Law of Treaties between States and International Organizations or between International Organizations, 1986. See Article 26 (*Pacta sunt servanda*):
Every treaty in force is binding upon the parties to it and must be performed by them in good faith.
And Article 27:
1. A State party to a treaty may not invoke the provisions of its internal law as justification for its failure to perform the treaty.

5

The Space Arms Race: Domain Asia

Ji Yeon-jung

The slow treaty-building process for the prevention of a space arms race has paradoxically unleashed an arms race in Asia. The emerging space powers in Asia – such as China, India, Japan, North and South Korea – appeared to learn from the nuclear arms race history: the application of armed space power might only be indeclinable if their space science and technology efforts reach an irreversible stage. The competition among Asian participants has become more intense than ever, based on the unrelenting motivation of self-interest backed by growing economic size. Despite each state's compliance with the Outer Space Treaty (OST), other related treaties, and the persistence of resolutions in the United Nations (UN) to uphold the peaceful use of outer space, their technological quest is contrary to these aims. The observable development of space technologies by the Asian states signal an impending militarisation and weaponisation of space. Thus, this chapter observes developments in Asian countries towards space arms race in the regional complex and their participation for the prevention of an arms race in outer space.

The Space Arms Race in Asia

In August 2016, China consolidated its leading status as a third space power by launching the world's first quantum communication satellite. The hack-proof communication satellite's design features the transmission of encrypted code detectable and alerted many contenders to the China-led competition to foolproof communication satellites against eavesdroppers. Many have interpreted China's development of this quantum communication satellite as a passive military asset, and it is assumed that this recent success would enhance China's confidence while also buttressing its conventional and nuclear capability and efforts to perpetrate cyberespionage. More importantly, it corroborates once again that China has morphed itself into a leading space power.

China's military intentions in outer space were duly noticed in 2005 and 2006 when it seemingly achieved the actual kinetic destruction of a satellite by successfully executing an unwarned launch of a "ground-based interceptor system into space", a proximity test to check the system's radar seeker.[1] The debates about China's military intentions gained further momentum with China's first successful anti-satellite test (ASAT) in addition to a two-stage mobile-launched missile, Dong Feng-21 (DF-21), in 2007. Still, China's space budget was less than 10 percent of the US's at that time. This, however, strengthened confidence in the cost-effective approach of China's space programme itself.[2]

In comparison to China's relatively late entry into the space race with its first satellite launch, Dong Fang Hong-1 (DFH-1), in 1970, Beijing's strategic calculation to advance into outer space was never in doubt. China's initial motivation for a space arms race was a by-product of the Cold War, as part of its challenge to confront the superpowers in terms of technological achievement. However, Beijing's paid a lot of attention to the use of US space assets in the execution of the First and Second Gulf War in 1991 and 2003, respectively. China's imaginings of space war games were also heightened when the US held a space war game scenario, SCHRIEVER I, designed to deploy US forces against the adversary-threatening allied island country, Taiwan. China's ASAT programme, for example, emerged as part of an Anti Access Area Denial (A2/AD) strategy against the US. As Johnson-Freese argues, the Chinese stepped into the arms race in response to the US. A Chinese official appearing in an interview with the Hong Kong News Agency said, "For countries that can never win a war with the United States by using the method of tanks and planes, attacking an American space system may be an irresistible and most tempting choice."[3] Since then, Beijing has prioritised space policy in the calculation of political and military hedges against its key adversaries – the US and other East Asian states – in the long-term.[4]

To be a third comer, China opted to take specific approaches that helped it to envision an effective pursuit of military space missions duplicating US efforts; this included missions' choice and concentration, and cost-effectiveness due to the countries' asymmetrical economies. China's priority on communication and reconnaissance satellites, for example, ascribes utmost importance to completing its defence plan along with the development of space weapons.

Largely, China envisioned completing its ASAT weapons capability, including direct-ascent anti-satellite missiles such as the SC-19 and DN-2, which are capable of destroying or disabling pre-selected satellites or spacecraft. In addition to completing a multi-dimensional programme, China is improving other technologies and concepts for directed energy or kinetic energy weapons for ASAT missions. The 2015 Report to Congress by the US-China Economic and Security Review Commission analysed China's increasing co-orbital proximity capabilities, noting that they include "jammers, robotic arms, kinetic kill vehicles, and lasers" to serve counter-space missions, even though "it may not develop or operationally deploy" those technologies.[5] However, the security dilemma drives all players into competition.

In the matrix of Asian geopolitics, India's pursuit of a space arms race was triggered by China's ASAT test.[6] India's responsive action was disclosed by General Deepak Kapoor, then Army Chief of Staff, in 2008, when he argued that India should "optimise space application for military purposes" corresponding to China's expanding space military programme, which is both offensive and defensive in nature.[7] His argument subsequently materialised when India's policy document, the Technology Perspective and Capability Roadmap, dictated India's initiative to develop ASAT weapons aiming at the "electronic or physical destruction of satellites in both the LEO [low earth orbit]- and GEO [geo-synchronous orbit]-synchronous orbits".[8]

The Defence Research and Development Organisation (DRDO) hastened the ASAT programme, especially after it successfully launched a 5,000-km range Agni V missile in 2012 capable of flying up to 600 km into space during its trajectory. Closer to developing ASAT weapons capability, then DRDO Chief V.K. Saraswat mentioned India's plan for an ASAT weapon that combined the Agni V's propulsion system with a "kill vehicle" as part of India's planned two-tier ballistic missile defence system, which was already in the pipeline.[9] With the increasing number of satellites being launched, India turned seriously towards protecting its space assets against adversarial countries. These efforts are also expected to serve as a consolidation of India's nuclear deterrence posture.

In developing the Agni V's capabilities, India also launched a radar reconnaissance satellite, the Radar Imaging Satellite-2 (RISAT-2), in 2009. When the next version of the satellite, RISAT-1, was launched in 2012, it provided a spatial resolution of approximately one meter in a polar sun-synchronous orbit. China identified India's spy satellite as being capable of succeeding the mission[10] India's targeted launching programme enables the country to launch satellites weighing more than 2 tons, and it is widely believed that India is technologically capable of deploying a weapons system into orbit.[11] Despite achieving these capabilities, India appears to be worried about China targeting its heavy communication satellite, which is comparatively slowly moving, easily detectable and targetable.[12] Like Vaydeesh Mahajan has warned, China's SC-19 DA ASAT weapon, using a mobile transporter-erector-launcher at Xichang, has in the past destroyed a satellite orbiting at an altitude of 860 km, so it might one day be a possible attacker on an Indian space asset.

However, it is not clear whether India will utilise its developing offensive space asset. A number of questions remain unanswered: How would India utilise ASAT capability in its war plans with China? How would India use ASAT tests in response to an adversary's attack, or would it seek to neutralise the adversary's capability in advance? Will India finally conduct an ASAT test when it is ready? How will India deal with the problems of debris from ASAT tests? In following the trend that a space treaty regime would be more restrictive in preventing the weaponisation of space, domestic concerns include the timing of India's successful technological demonstration must collaborate with India's envisioned status in legal regime in the field of the outer space.

Unlike the patterns of China and India in the space arms race, Japan has formulated a more cautious approach; yet it is moving steadily on the track. Similar to China, the Japanese Government also began to study the feasibility of creating a reconnaissance satellite programme in 1991, primarily inspired by the use of space programme in US military operations.[13] One of the Japan's intentions for the space race correspondingly influenced in speculation of an extreme case; uncooperative US-Japan relations. Japan's over-reliance on US military space assets and the consequent vulnerability is a further reason for Japan's venturing into the space military purposes. As Tao Nakayama, then Japanese Foreign Minister, stressed to the Foreign Affairs Committee of the Japanese House of Representatives the necessity of an intelligence satellite programme: "If we don't receive intelligence from America, we won't know anything."[14] His opinion was supported by many political leaders and soon materialised into a practical study conducted by the Defence Policy Bureau, Defence Agency, in response to North Korea's Nodong missile test in 1993. One conclusion drawn from this panel meeting called for the Japanese Government to develop an Earth-observation satellite.[15]

Information supremacy then appeared to be a priority in Japan's space policy and indicated Japan's foreseeable entry into the space arms race with mature launching technology which can be converted into missile knowhow. Japan launched a military reconnaissance satellite, the Information Gathering Satellite (IGS), in 2003.[16]

However, Japan's 2003 IGS launch elicit regional player's suspicion. Japan's possible entry into the space arms race was constrained by a Diet resolution in 1969 that capped its military operational actions in space. However, the original guidelines for the peaceful use of space as a non-military venue were re-interpreted in the resolution as "the 'non-aggressive' military use of space".[17] This move was discernible when the Liberal Democratic Party introduced the Basic Space Bill to the House of Representatives in 2007, and with its subsequent passage in August 2008. The Basic Space Law lifted a ban that dictated that space should not be used for defensive purposes. Since 2009, there has been a growing public and industrial demand to unleash Japan's self-restriction on space-based defence, witnessed by Japan's steady turn to the space arms race with full public awareness. In 2009, the Japanese Government's efforts gained momentum after the launch of the military reconnaissance satellite IGS-Optical 1.

In 2016, Japan agreed to allow Taiwan to use Japanese spy satellites until Taiwan's surveillance satellite, Formosat 5, could be launched. Japan's tacit collaboration with Taiwan since 2014 has led to the assumption that two states share a convergence of interests in managing the regional political complex and rivalries.[18] While Japan's threat perception officially emphasises North Korea's missile launches, the actual space strategy seems calibrated to counter China.

Rather North Korea's nuclear aggression and ballistic missile tests invite South Korea's entry into military satellite programme. Until the early 1990s, South Korea had no organisational structure or plan to join the space race. In response, it has exerted great effort to attempt "fast-track development" and to implement a more integrative strategy to overcome its more comparatively limited financial and human resources.[19] South Korea launched the multipurpose satellites, Arirang 1 and 2 in 1999 and 2006, respectively, and the latter launch made South Korea the seventh country to possess a

1m-class high-resolution satellite. However, it was the Arirang 3 (KOMPSat 3), launched in 2012, that elevated Seoul to the level of an emerging player. Primarily, the South Korean military has pushed to mobilize its intelligence capability to monitor enemies' military activities. North Korea's series of nuclear and ballistic missile tests has prompted Seoul's intent to boost its projects and initiated an ambitious plan to place five military satellites in orbit by 2022.[20]

While the South's predominant target is the deterrence of North Korean aggression, Seoul is aware that future warfare will be in the domain of outer space, and regional competition has intensified.[21] To meet the demands of future warfare needs, Seoul plans to build its defensive space power in a multi-phase manner, as first announced by the Ministry of National Defence in 2012. Simultaneously, in 2014, South Korea amended the Space Development Promotion Act of 2005 to fill the lacuna of space development corresponding to the requirement for national security.[22]

Overall, all major players in Asia are on the path to militarisation of space; they seek to extend their technological reach and their legal basis for doing so. As the speed of China's efforts in the space arms accelerate, other regional players will further expand their space programmes and infrastructure to attain international competitiveness.

Arms Control in Outer Space

The leading space players in Asia are also active participants in the process of building a space arms control regime. Four countries – China, Japan, India, and South Korea – are signatories of the OST, yet they have not signed the Moon Agreement, the treaty not to militarise the Moon and other celestial bodies. Moreover, little coordination is visible among the Asian countries when it comes to defining and negotiating legal and normative structures for an arms control regime in space. China, foremost, has extended its diplomatic theatre to the global stage, neutralising the US's refusal to discuss arms control in outer space with Russia. China's continuing request to hold talks on the proposed agreement towards the prevention of an arms race in outer space (PAROS) since 2002 has witnessed Beijing's diplomatic activeness towards space treaty regime building.

On October 27, 2014, Shen Jian, counsellor of the Chinese Delegation at the Thematic Debate on Outer Space at the 69th session of the United Nations General Assembly (UNGA), highlighted China's dedication to working on a multilateral agreement that prevents arms races in outer space and promotes transparency and confidence-building measures (TCBM). Along with Russia, China submitted a revised draft of a working paper titled, "Treaty on the Prevention of the Placement of Weapons in Outer Space, the Threat or Use of Force against Outer Space Objects (PPWT)" which unceasingly confronts US obliviousness. The key suggestions of the PPWT are not to place "any" weapons in outer space, not to resort to the threat or use of force against state parties' objections, and not to engage in outer space activities against the PPWT or assist or induce any state, non-state, intergovernmental actors, and so on to join in activities inconsistent with the PPWT.[23] China's developing work contributes to promoting awareness on the space arms race; however, it still fails to address many critical issues such as defining terms of the proposed treaty, improving the verification process, managing space debris and suppressing ground-launched weapons that target orbital assets.[24]

On the surface, it appears that China's primary intention is to prevent a competitive space arms race. However, China's strategic calculation are inspired by the need to tackle US space supremacy, especially on competitively successful missile defence and more operational space weapons. The technological significance also meets the political desire to build a China-led international consensus to prevent an arms race vis-à-vis the Code of Conduct for Outer Space Activities suggested by the European Union (EU) and backed by the US-China's continued involvement in these negotiations attests to its interest in space arms control regime.

China's proactive participation in this fields invites a complicated conflict of interest among regional players. India that holds an active historical record in nuclear arms control negotiation is wary of China's expanding diplomatic influence. D.R. Venkatesh Varma, ambassador and permanent representative of India to the Conference on Disarmament, has mentioned India's approach and activity in discussing the EU-led draft of the international Code of Conduct for Outer Space and its entry into the United Nations (UN) Group of Government Experts (GGE) on the TCBMs.[25] One of India's major concerns appears to be whether to hold a leading position in the various international negotiations of the outer space domain, thus avoiding technological isolation.[26] India's experience since 1974 on inaccessibility to space technology transfer keeps New Delhi attentive to exploring all ways to join the space treaty regime, possibly as a leading country equivalent to the permanent five members on the UN Security Council, or at least as a member of the treaties. India's aspiration of becoming a member of the space treaty regime is a part of a larger diplomatic design to enter the nuclear non-proliferation regime and disarmament process.

Tokyo also is moving forward to continue work on the Space Security Conference in the United Nations Institute for Disarmament Research and GGE on TCBMs, in tandem with its favourable approach to the EU-proposed draft of the Code of Conduct.[27] Japan's approach to space arms control plays within disarmament policy yet of US-Japan space collaboration in the context of the alliance. In March 2013, the US and Japan agreed on a legal framework for the "provision of US space situational awareness data" that further accommodated a bilateral partnership to share a geographical advantage across Asia.[28] Japan acknowledged that the present legal framework does not categorically prohibit the test or use of ASAT weapons, which Asian faring states pursue.[29] The geographical proximity and regional rivalry steers Japan to be more vigilant on China's ASAT weapons capability than others'. The strategic vulnerability in outer space elicits stronger governmental diplomacy to prevent an arms race in space yet a paradoxical reverse to unleash self-restriction on military space missions.

The complicated regional competition entangled with the immediate threat perception hinders regional space cooperation by extension. Whereas international space cooperation in preventing the space arms race is prioritised foremost in Asian space-faring states, the dyadic rivalry and military framework in alliances receive benign support among competitors. China's Asia-Pacific Space Cooperation Organisation (APSCO), evolved from the Asia-Pacific Multilateral Cooperation in Space Technology and Applications (AP-MCSTA), halfway succeeded to be the leading regional forum to institutionalise a space cooperation mechanism. The nascent plan of AP-MCSTA was convened with three parties, China, Pakistan, and Thailand, that subsequently expanded with six other countries. However, the emerging Asian space powers least participate in this institute. The Japan-led Asia-Pacific Regional Space Agency Forum (APRSAF) is also unable to gain China's support despite the growing number of participating countries. Among members, technical and educational components are main pillars of space cooperation in APRSAF that exclude technology transfer and security-related dialogue. An inherent limitation of APRSAF, however, not only lies in regional rivalry, but also constraints due to the lack of regional legal mechanisms for technology transfer.

Conclusion

Space remains in the domain of the future. The major Asian space aspirants strive to attain technological parity and international prestige to capture a larger share of outer space. Since future political and military confrontations inevitably extend to outer space, space security has become a pillar of national security. In Asia, regional rivalries encourage boundless competition for the militarisation and weaponisation of outer space despite independent participation in establishing a universal space treaty regime to prevent an arms race. To speed up the international process, regional support from emerging space powers is indispensable. A lack of transparency and competition regarding a leading role in the space treaty regime can only be solved in building trilateral or quadrilateral talks among China, India, Japan and South Korea, even with trivial initiation of an exchange of opinions on the creation of a regional forum, varied regional opinions on draft treaties to prevent an arms race, and consensus to provide the collective security of commercial satellites.

ENDNOTES

1. James Clay Moltz, *Asia's Space Race: National Motivation, Regional Rivalries, and International Risks*, Columbia University Press, New York, 2012, p.93.
2. "Official Says China Spends Less than 10 per cent of US Space Budget", BBC Monitoring Asia Pacific, October 16, 2006, p. 1.
3. Joan Johnson-Freese, "Space Wei Qi: The launch of Shenzhou V", *Naval War College Review*, LVIL (2), 2004, p. 129.
4. Erik Seehouse, *The New Space Race: China vs. the United States*, Praxis Publishing, Chichester, 2010, p. 40.
5. "2015 Report to Congress of the U.S.-China Economic and Security Review Commission, Executive Summary and Recommendations", One Hundred Fourteenth Congress, First Session, November 2015, at http://origin.www.uscc.gov/sites/default/files/annual_reports/2015%20Executive%20Summary%20and%20Recommendations.pdf
6. Rajeswari Pillai Rajagopalan, "India's Changing Policy on Space Militarization: The impact of China's ASAT Test", *India Review*, 10 (4), 2011, pp. 354-378.
7. Peter Hulsroj, Spyros Pagkraitis and Blandina Baranes (eds.), *Yearbook on space policy 2010/2011: The Forward Look*, Springer, Vienna, 2013, p. 73. David R. Sands, "China, India Hasten Arms Race in Space", *The Washington Times*, June 25, 2008, at http://www.washingtontimes.com/news/2008/jun/25/china-india-hasten-arms-race-in-space/
8. Rajeswari Pillai Rajagopalan, No. 6.
9. Rajat Pandit, "After Agni-V Launch, DRDO's New Target is Anti-Satellite Weapons", *The Times of India*, April 21, 2012, at http://timesofindia.indiatimes.com/india/After-Agni-V-launch-DRDOs-new-target-is-anti-satellite-weapons/articleshow/12763074.cms
10. "India Successfully Launches Spy Satellite RISAT-1", *Xinhua*, April 26, 2012, at http://news.xinhuanet.com/english/world/2012-04/26/c_131552885.htm.
11. Harsh Vasani, "India's Anti-Satellite Weapons", *The Diplomat*, June 14, 2016, at http://thediplomat.com/2016/06/indias-anti-satellite-weapons/.

12. Vaydeesh Mahajan, "Chinese Anti-Satellite Means: Criticality and Vulnerability of Indian Satellite", *CLAWS Journal*, Summer 2016, p. 183.
13. William W. Radcliffe, "Origins and Current State of Japan's Reconnaissance Satellite Program (U)", *Studies in Intelligence*, 54 (3), p. 11.
14. Ibid.
15. James Clay Moltz, No. 1, p. 57.
16. Kim Sung Chull, "Japan's Strategic Response to North Korea: Activistic Security Policy, Eroding Pacifism", in Purnendra Jain and Peng Er Lam (eds.), *Japan's Strategic Challenges in a Changing Regional Environment*, World Scientific Publishing, Singapore, 2013, p. 79.
17. Setsuko Aoki, "Japanese Perspectives on Space Security", in John M. Logsdon and James Clay Moltz (eds.), *Collective Security in Space: Asian Perspectives*, George Washington University, Washington DC, 2008, p. 62. Christopher W. Hughes, "Japan's Military Modernisation: A Quiet Japan-China Arms Race and Global Power Projection, *Asia-Pacific Review*, 16 (1), 2009, p. 10.
18. "Space: Taiwan and Japan in the Same Orbit", *Strategy Page*, July 30, 2016, at https://www.strategypage.com/htmw/htspace/articles/20160730.aspx.
19. James Clay Moltz, No. 1, p. 137.
20. "Korea to Resume Project to Launch Five Military Satellite", *The Korea Times*, March 31, 2016, at http://www.koreatimes.co.kr/www/news/nation/2016/03/116_201627.html.
21. "2014 White Paper", Ministry of National Defense, Republic of Korea, pp. 13, 15, at http://www.mnd.go.kr/user/mnd_eng/upload/pblictn/PBLICTNEBOOK_201506161152304650.pdf.
22. Ibid. p.64.
23. Letter dated June 10, 2014, from the permanent representative of the Russian Federation and the permanent representative of China to the Conference on Disarmament, addressed to the acting secretary-general of the Conference, transmitting the updated Russian and Chinese texts of the draft of the "Treaty on the Prevention of the Placement of Weapons in Outer Space, the Threat or Use of Force against Outer Space Objects (PPWT)", introduced by the Russian Federation and China, Conference on Disarmament (CD), CD/1985, June 12, 2014, at https://documents-dds-ny.un.org/doc/UNDOC/GEN/G14/050/66/PDF/G1405066.pdf?OpenElement.
24. Michael Listner and Rajeswari Pillai Rajagopalan, "The 2014 PPWT: A New Draft But with the Same and Different Problems", *The Space Review*, August 11, 2014, at http://www.thespacereview.com/article/2575/1.
25. "Statement by Mr. D.B. Venkatesh Varma, Ambassador and Permanent Representative of India to the Conference on Disarmament at the CD's Informal Plenary on Agenda Item 3 on the Prevention on an Arms Race in Outer Space (PAROS)", June 11, 2014, athttp://www.mea.gov.in/Portal/CountryNews/2691_PR_Statement_on_PAROS_june_2014.docx
26. Ajey Lele, *Asian Space Race: Rhetoric or Reality*, Springer, New Delhi, 2013.
27. "Statement by Mr. Hiroyuki Yamamoto, Minister, Delegation of Japan to the Conference on Disarmament, 'Prevention of an Arms Race in Outer Space'," 1283rd Plenary Meeting of the Conference on Disarmament, Geneva, March 19, 2013, Government of Japan, at http://www.disarm.emb-japan.go.jp/Statements/130319CD.htm.
28. *Space Cooperation: Space Situational Awareness Service and Information, Agreement between the United States of America and Japan, Effected by Exchange of Notes at Tokyo May 28, 2013*, Treaties and Other International Acts Series 13-528, Government of the United States.
29. Setsuko Aoki, No. 17, p. 64.

SECTION II

GLOBAL OUTLOOK

6

The European Union and the Outer Space Treaty: Will the Twain Ever Meet?

Frans G. von der Dunk

Introduction

In spite of the envisaged Brexit and other crises and problems currently threatening the European Union (EU)[1], that half-way house between a group of cooperating states and a single quasi-federal union of states remains an important player in today's world, also – at least from a bird's eye view – in terms of outer space. Its member states Germany and France have the largest space budgets of all European states (discounting the Russian Federation as a European state), and the European flagship projects Galileo and Copernicus, with the European Commission on behalf of the Union in the driver's seat, are among the most challenging and interesting space infrastructures currently being developed.

That, obviously, then also raises the issue of the EU's 'relationship' with, views on and involvement with the Outer Space Treaty[2], the most comprehensive and generic international convention setting out the legal framework for all space activities. It should be noted at the outset that the Treaty itself, drafted in the middle of the Cold War and focusing on military and scientific aspects of space activities, is very much targeting its legal regime towards sovereign states, not towards a unique 'phenomenon' such as the Union.[3]

Prior to going into the specifics of any 'relationship' of the EU with, views on and involvement with the Outer Space Treaty, however, it is important to understand the proper place of the Union and its predecessor in the broader European space endeavour – as it is by no means the only, or even the first European body within the European 'spacescape'.

The Broader European 'spacescape'

Actually, the first such European bodies were the European Organisation for the Development and Construction of Space Vehicle Launchers (ELDO)[4], established in 1962 to develop a joint European launcher after neither the UK nor France had been able to pull off a viable launcher development programme on their own, and the European Space Research Organisation (ESRO)[5], established the same year to coordinate and integrate (some of) the space research programmes of the individual member states – a group largely overlapping, but not completely identical to that of ELDO member states.[6]

When both organisations turned out to fall far short of their intended goals, in 1975 it was decided to essentially merge the two into one: thus, the European Space Agency (ESA) was established.[7] Australia left the club that instant; ESA has since then grown to encompass 22 European-only member states: Austria, Belgium, the Czech Republic, Denmark, Estonia, Finland, France, Germany, Greece, Hungary, Ireland, Italy, Luxembourg, the Netherlands, Norway, Poland, Portugal, Romania, Spain, Sweden, Switzerland and the UK.[8]

ESA soon covered almost the complete spectrum of the joint European space effort in terms of types of activities, from pure space science to prototype communication and earth observation satellites. The boundaries of where ESA was deemed by the still-sovereign member states – of which especially the larger ones also maintained their own national space programmes – to provide the best possible venue for their space efforts were set by the ESA Convention:

The purpose of the Agency shall be to provide for and to promote, for exclusively peaceful purposes, cooperation among European States *in space research and technology and their space applications, with a view to their being used for scientific purposes and for operational space applications systems,*

(a) by elaborating and implementing a long-term European space policy, by recommending space objectives to the Member States, and by concerting the policies of the Member States with respect to other national and international organizations and institutions;

(b) by elaborating and implementing activities and programmes in the space field;
(c) by coordinating the European space programme and national programmes, and by integrating the latter progressively and as completely as possible into the European space programme, in particular as regards the development of applications satellites;
(d) by elaborating and implementing the industrial policy appropriate to its programme and by recommending a coherent industrial policy to the Member States.[9]

In other words, as soon as certain prototype technologies had proven their feasibility, it was not deemed appropriate for ESA to continue operating them on a 'routine' basis for downstream terrestrial applications. Thus, once ESA had developed an operational launcher, in 1980 its member states created Arianespace as a private French company with involvement and support from about half of the current ESA member states to operate these launches on a commercial basis.[10] Once ESA had demonstrated the success and operational viability of the satellite communications technology it had developed, in 1982 EUTELSAT was established – with ultimately more than double the amount of member states compared to ESA (but all European) – to operate a European satellite infrastructure for communication purposes on a quasi-commercial basis.[11] Similarly, in 1983 its member states created EUMETSAT to run with the earth observation satellite technology ESA had developed and provide earth observation services for meteorological, then also climate change purposes to its member states – again, a set different (in this case slightly larger) from the membership of ESA itself.[12]

As it is yet too early for Europe's satellite navigation system Galileo (developed, as far as the technology goes, under ESA auspices)[13] or Europe's environmental and security monitoring system Copernicus (equally being developed, as far as the technology goes, under ESA auspices)[14] to be operated on a daily 'routine', let alone commercial basis, it remains unclear for the time being what specific respective governance systems are to be established for the long term.

This finally brings us to the position of the EU within this larger 'European spacescape', as it also constitutes the main reason why the governance systems for these two 'European flagship projects' will most likely not consist of a different and formally independent entity such as Arianespace, EUTELSAT/Eutelsat or EUMETSAT, but of an EU body or agency.[15] In many respects, this driving role for the EU represents the culmination of three decades of growing interest and involvement of the Union and its predecessor, the European Community, in outer space and space activities.

The European Communities/Community/Union enters the European 'spacescape'

For a proper understanding of why the European Communities for a long time did not play any role in the European 'spacescape' or vice versa, one has to understand furthermore the background to and rationale of that particular vehicle for European cooperation. From the start in the 1950s, three intergovernmental organisations of a very special nature – the European Coal and Steel Community (ECSC)[16], within a few years followed by the European Atomic Energy Community (EAEC, also Euratom)[17] and the European Economic Community (EEC)[18] with identical sets of member states – were established in Europe *for the purpose of regulating trade across member state borders on the basis of an Internal Market with considerable socio-economic and political safeguards.*

Institutionally speaking, their uniqueness transpired in the way several Community-level bodies – the predecessors of the current European Commission, Council of Ministers, European Parliament and Court of Justice of the European Communities – were able to take decisions binding upon the totality of member states and their citizens in the respective realms addressed by the treaties, overriding as necessary any diverging national legislation on the subject.[19]

Substantively speaking, where the ECSC and EAEC by definition confined themselves to specifically circumscribed and quite special areas of the European economies, the EEC was supposed to address the remainder of economic activities – *to the extent addressed, explicitly or implicitly, by the EEC Treaty or implementing European level-regulation, which meant in essence as soon as a certain sector of the economy had become commercialised, privatised and subject to substantial international trade* and *the European institutions had officially recognised this.*

This is the key: until the mid-1980s, there simply was no space sector in Europe which fitted those conditions, hence 'space' figured only in some visionary political documents, but neither in any legislative activities nor in any material actions. Neither the growth of the three Communities in terms of member states nor the 'institutional integration' thereof which took place through various later treaties ultimately giving rise to the European Community and the EU,[20] changed this fact.

In the mid-1980s, however, two factors changed this fundamental attitude of abstinence.[21] On the one hand, the EEC became increasingly pro-active in stimulating the European economy (read the totality of economies of the member states) not only by paving the way in regulatory terms for free trade and commerce, but also by supporting research and development as far as acting as catalysts to economic growth. The 1986 Single European Act thus for the first time gave the EEC such a pro-active role in research and development – which was generally agreed to including space-related research and development.[22]

On the other hand, even more importantly, the mid-1980s can be seen as heralding the commercial viability of satellite communications, the first sector of space with clear commercial benefits, also in Europe, with the establishment of the private satellite operator SES in Luxembourg in 1985 to start competing with the then-still-intergovernmental EUTELSAT. The result was a first major piece of EC legislation on space, notably laying the foundations for a single European market for the provision of satellite services, by way of the so-called 'Satellite Directive' in 1994.[23]

On-going market developments in the area of satellite communications led the European Commission also to force EUTELSAT to privatise;[24] in addition, in certain other areas the European authorities also started to enter the European 'spacescape' with efforts to regulate certain commercial or commercially-relevant aspects thereof. Most notable was the so-called 'Database Directive'[25] of 1996 which established a *sui generis* copyright for electronic databases prominently including satellite remote sensing databases, since existing copyright did not sufficiently cover such novel phenomena – and protection of intellectual property rights was considered key for developing a viable commercial remote sensing sector.

With the increasing realisation that the space sector, as a high-key technology sector, became key to general European economic development as well, the European Commission's approach to 'outer space' and 'space activities' grew progressively more comprehensive, although always limited by the extent to which the European treaties and member states allowed.[26] In practical/political terms moreover such efforts also faced the existing realities of a number of entrenched European entities being involved in the European 'spacescape', most prominently of course ESA.

The two European Flagship Projects: Galileo and Copernicus

The two European flagship projects which were initiated by the European

Commission in the late 1990s/early 2000s not only represented the culmination up to that point of increasing EU involvement in space activities, it also presented a watershed. In line with the general role of the Union as a (quasi-)legislative machinery at a partly supra-national level, until then such involvement (along the lines sketched above) had taken place in the politico-legal area, by way of enunciation of certain Directives and Regulations mainly addressing market and research-and-development aspects of certain space activities or applications.

With Galileo, however, the Union for the first time started to become an actual 'space player' itself. Following initial efforts to become partners with the US in operating Global Positioning System (GPS) as a Global Navigation Satellite System (GNSS), the Commission decided it would be in the European strategic, political and economic interest to develop a separate and independent European satellite navigation system, which came to be called 'Galileo', to be built upon a satellite navigation augmentation system still working with GPS called the European Geostationary Navigation Overlay Service (EGNOS) developed prior to Galileo-proper.[27]

Through a small series of EU legislative documents[28] the Commission increasingly took control over the development and deployment phases of the EGNOS and Galileo programmes, especially after the original intention to have a concessionaire take over the system and operate it on a commercial basis had fallen through. The role of ESA, originally almost on a par with that of the Union/Commission as providing the necessary technological know-how, correspondingly receded into the background; by now it has essentially been relegated to the role of procuring agency on behalf of the Commission.[29] In particular by owning the satellites comprising the space-part of the system, both as launched and as yet to be launched,[30] the EU may now be said to have become a space operator at least in a formal-legal sense.

Following upon the initial positive reaction within the EU member states to Galileo, at least on the strategic-political level, the Commission felt emboldened to then also develop plans for an independent European satellite infrastructure for earth observation purposes, more specifically for Global Monitoring for Environment and Security (GMES).[31] Meanwhile rechristened 'Copernicus', following three Regulations[32] the Commission had been in the driver's seat from the very beginning, with ESA as a kind of junior partner[33] currently responsible for the few Sentinel satellites that already have been launched.

At a later stage, although the details of the institutionalisation are far from clear and as of yet only a 'GMES Bureau' has been created within the Commission to prepare such a future governance regime, it may be assumed that the Union will, along roughly similar lines as with Galileo, become the ultimate responsible authority for the space activities conducted by and with the Copernicus infrastructure.[34]

In the context both of Galileo and of Copernicus, the fact that two of the participating countries – Norway and Switzerland – are member states of ESA only, thus co-financing the programmes to the extent that ESA is executing them, yet not member states of the EU, so not automatically involved in the decision-making on the future of both programmes, has so far been solved pragmatically. Yet it has to be kept in mind – as this membership-situation is not likely to change anytime soon – that this may still come to present problems in the future.

The Final Step: Comprehensive EU Competence in Space?

Largely at the background of the developments mentioned earlier, a more long-term political discussion took place on the future of European space policy and law, and in particular the overall relationship between ESA, the erstwhile vehicle for all such discussions and still equipped with the necessary technical and operational know-how, and the Union/Commission, with clearly superior powers in the political and legislative realm but, as seen, a relatively late entrant into the European 'spacescape'. While the Commission originally tended to go for incorporation of ESA into the Union-structure, as a 'space agency of the Union',[35] already the different memberships – currently, amongst the 28 EU member states and 22 ESA member states, 20 countries belong to both groups – precluded that from happening in the short run.

Nevertheless, the two entities willy-nilly converged in their operations, approaches and activities in the European 'spacescape'.[36] Still largely on the basis of 'equality' the two concluded a Framework Agreement in 2003[37], cementing their overall cooperation while leaving each to do what it was best at – in accordance with its own internal procedures and *modi operandi*.

The major development here concerned the development of an overall EU competence in space as opposed to the isolated areas where it had already exercised its legislative authority on the basis of Internal Market competences.[38] The end result so far is the inclusion, following the Treaty of Lisbon[39], in the Treaty on the Functioning of the EU of an Article providing the following:

1. To promote scientific and technical progress, industrial competitiveness and the implementation of its policies, the Union

shall draw up a European space policy. To this end, it may promote joint initiatives, support research and technological development and coordinate the efforts needed for the exploration and exploitation of space.

2. To contribute to attaining the objectives referred to in paragraph 1, the European Parliament and the Council, acting in accordance with the ordinary legislative procedure, shall establish the necessary measures, which may take the form of a European space programme, *excluding any harmonisation of the laws and regulations of the Member States.*
3. The Union shall establish any appropriate relations with the European Space Agency.
4. This Article shall be without prejudice to the other provisions of this Title.[40]

In particular because of the italicised phrase, this 'space competence' has been labelled a "parallel competence", as it – contrary to the 'standard' approach of EU law – leaves the authority of individual EU member states to enunciate domestic legislation in tact.[41] Recognising the fact that currently seven out of 28 EU member states have a comprehensive national space law providing for a licensing regime of private space activities (in chronological order: Sweden[42], the UK[43], Belgium[44], the Netherlands[45], France[46], Austria[47] and Denmark[48]), this quite seriously limits the actual possibility for the Union to draft overarching EU law in this particular context.[49] It thus still remains to be seen to what extent the Union will be allowed to further develop its legislative role in the field of European space activities and applications.

The above analyses clarify that in the end the Union's relationship with the Outer Space Treaty is essentially twofold, noting that of the 28 EU member states four are not parties to the Treaty (Croatia, Latvia, Malta and Slovenia) but that the provisions are generally recognised to present customary international law.

The EU as a 'legislator' and the Outer Space Treaty

On the one hand, to the extent the Union acts in its legislative capacity, whatever law or regulation it enacts, such law or regulation should not run counter to the provisions of the Outer Space Treaty – and at least 24 out of 28 member states are bound to ensure the Union does not do so, and are also bound to succeed in doing so in view of their large majority. The Union itself also recognises the fundamental obligation resting upon it to comply with international law, and as the Outer Space Treaty reflects customary international law, this clearly includes that Treaty as well, even as the Union itself has not formally indicated this specifically.[50]

However, the provisions of the Outer Space Treaty are rather broad and general, providing only general obligations to:

1. Act in the interests of the international community, international cooperation and international peace and security.[51]
2. Desist from the stationing of weapons of mass destruction in outer space.[52]
3. Treat astronauts as 'envoys of mankind' and support them when in distress as much as possible.[53]
4. Accept international responsibility and liability as appropriate, and ensure proper authorisation and continuing supervision of non-governmental space activities.[54]
5. Desist at least in principle from harmful interference with other (States') legitimate space activities.[55]
6. Allow access in principle to stations and equipment on celestial bodies.[56]
7. Generally comply with international law applicable to outer space.[57]

While the very last obligation mentioned also raises the issue of the extent to which *other* space or space-related treaties, even if not specifically mentioned or ratified, would as extensions or elaborations of the Outer Space Treaty, also have to be complied with, at this stage it suffices to note that so far all EU legislation addressing space activities or issues has been fairly limited and rather focused on specific aspects. Therefore, it can safely be said that this overarching requirement of EU compliance with the Outer Space Treaty is complied with so far – and until the 'space competence' of the Union pursuant to the Treaty of Lisbon would really come to be exercised, that is not likely to change.

The most pertinent example here concerns the satellite communications realm, where, as indicated, the Union is in the process of harmonising the market conditions within the Internal Market ever since the 1994 Satellite Directive. Since commercial uses of outer space at least in the realm of satellite communications, as long as compliant with the regime developed in the framework of the International Telecommunication Union (ITU) regarding the use of orbital slots/orbits and space frequencies[58], are undisputedly allowed, the main other overarching requirement stemming from the Outer Space Treaty concerns that of Article VI, to have such activities by private operators properly authorised and supervised by the appropriate state(s).[59] The regime imposed by the Union does exactly heed that requirement: only properly licensed satellite communication operators may enjoy the benefits of that Internal Market for telecommunication service providers, and though subject to some EU framework conditions, such licences are to be granted by national telecom authorities.

The EU as a 'space operator' and the Outer Space Treaty

On the other hand, to the extent the Union acts as a space operator – which in the case of Galileo and, at least soon, Copernicus, would ultimately seem to be the case – it obviously has to comply with the regime set out by the Outer Space Treaty as well. As, again, in substantive terms, clear-cut legal obligations are only found in a fairly limited context, the 'institutional place' of the Union in the framework of the Treaty is probably the most directly relevant aspect of the relationship of the former to the latter.

In the Treaty, namely, there are but two clauses making reference to international intergovernmental organisations. First, "[t]he provisions of this Treaty shall apply to the activities of States Parties to the Treaty (…) including cases where they are carried on within the framework of international intergovernmental organizations", and "[a]ny practical questions arising in connection with activities carried on by international intergovernmental organizations (…) shall be resolved by the States Parties to the Treaty either with the appropriate international organization or with one or more States members of that international organization, which are Parties to this Treaty".[60]

Second, "[w]hen activities are carried on in outer space (…) by an international organization, responsibility for compliance with this Treaty shall be borne both by the international organization and by the States Parties to the Treaty participating in such organization".[61] Thus, following the politico-legal logic of the Soviet Union at the time, which did not want to accord any special status to international organisations, ultimately such organisations remained, legally speaking, platforms for cooperation rather than independent legal persons capable of acting with the slightest independence from their own member states.[62]

This also means, that regardless of the 'space competence' and the extent to which it would allow for the Union to start encroaching upon the licensing regimes of the seven states which so far have a national space law in place, also from the vantage point of Article VI of the Outer Space Treaty EU-level authorisation and continuing supervision of private space activities is neither a requirement nor a right.

The major remaining issues would thus concern the obligations set out in their most fundamental version in Article VII (if a state is involved in the launching of a space object in any of the manners indicated, it will be held liable for damage caused by such a space object) and Article VIII (if a space object is launched into outer space, it is supposed to be registered by (one of) the state(s) involved, thus giving such state jurisdiction over the object). Since, however, the Union is not a state in any relevant legal sense of the word, as further supported by the provisions of Articles VI and XIII quoted above, there will be EU member states carrying such liabilities and enjoying the obligation-*cum*-possibility to register in its stead; how such liabilities, obligations and exercise of jurisdictions would then further be given shape is essentially an internal matter for the Union – so far not at all touched upon.

Will the Twain ever Meet?

The above analyses already make clear that, beyond the general obligations of the EU and its organs to stay in line with the Outer Space Treaty's regime, both in its role as legislator and in its role as space operator, there will be few direct connections between the Union and the Outer Space Treaty. Different from follow-on space treaties developed in the bosom of the United Nations, the Outer Space Treaty does not offer any intergovernmental organisation to act as a de factoparty to such a treaty. Neither is there any other category of legal subjects in the Outer Space Treaty that it could feasibly be included under. So effectively the space activities that are and will be undertaken under its aegis are to be accounted for by the individual EU member states also party to the Outer Space Treaty.

Where for instance Article XXII of the Liability Convention[63] and Article VII of the Registration Convention[64] do offer intergovernmental organisations the possibility to become a de factoparty to those respective treaties, the Union apparently does not see itself as an 'ordinary' intergovernmental organisation and thus is not prepared to make use of these possibilities.

So, will the twain ever really meet? To reiterate: the only two realistic options so far are for the Union to indeed start licensing private space operators directly so as to become subject to the rights and obligations of Article VI of the Outer Space Treaty (which still requires an argument that, apparently the solution found by the EU member states pursuant to Article XIII would require such direct EU licensing, allowing it to override any stricter interpretation that Article VI, really, only can address states full stop), or to accept a secondary status as an 'intergovernmental organisation' launching and registering space objects pursuant to the Liability Convention (and hence also subject to Article VII of the Outer Space Treaty) respectively the Registration Convention (and hence also subject to Article VIII of the Outer Space Treaty). The first goes against the current trend in the attitudes of the member states concerned; the second against the EU's own political approach – if indeed the twain will ever really meet, it will be at least a few years away.

ENDNOTES

1. As the actual, that is, legally-applicable Brexit, is at least a few years away (and might theoretically even turn into something different from a formal Brexit after all), for the purposes of this contribution, the UK will be considered part of the Union and the Union will be considered to have 28 member states.
2. Treaty on Principles Governing the Activities of States in the Exploration and Use of Outer Space, including the Moon and Other Celestial Bodies (hereafter Outer Space Treaty), London/Moscow/Washington, done January 27, 1967, entered into force October 10, 1967; 610 UNTS 205; TIAS 6347; 18 UST 2410; UKTS 1968 No. 10; Cmnd. 3198; ATS 1967 No. 24; 6 ILM 386 (1967).
3. See P. Jankowitsch, "The Background and History of Space Law", in F. G. von der Dunk and F. Tronchetti (eds.), *Handbook of Space Law*, Edward Elgar Publishing, Cheltenham/Northampton, 2015, pp. 1-14; M. Lachs, *The Law of Outer Space: An Experience in Contemporary Law-Making*, Brill | Nijhoff, 2010 (reprint), pp. 11-25; P. Malanczuk, "Space Law as a Branch of International Law", *Netherlands International Law Yearbook*, 25, 1994, pp. 143-80; N. M. Matte, "Space Law", in R. Bernhardt (ed.), *Encyclopedia of Public International Law* , Vol. 4, 2000, pp. 552-7; G. Zhukov and Y. Kolosov, *International Space Law*, Praeger Publishers, New York, 1984, pp. 1-17; B. Cheng, *Studies in International Space Law*, Clarendon Press, Oxford, 1997, esp. pp. 70-87.
4. ELDO was established by means of the Convention for the Establishment of a European Organisation for the Development and Construction of Space Vehicle Launchers, London, done March 29, 1962, entered into force February 29, 1964, expired October 30, 1980; 507 UNTS 177; UKTS 1964 No. 30; Cmnd. 2391; ATS 1964 No. 6. Amongst its member states, Australia as a non-European state stood out – the use of its launch site for European launches was considered to be of mutual interest to both the European member states and Australia itself.
5. ESRO was established by means of the Convention for the Establishment of a European Space Research Organisation, Paris, done June 14, 1962, entered into force March 20, 1964, expired October 30, 1980; 158 UNTS 35: UKTS 1964 No. 56; Cmnd. 2489.
6. ELDO comprised Australia, Belgium, France, Italy, the Netherlands, the UK and West Germany; whereas ESRO comprised Belgium, Denmark, France, Italy, the Netherlands, Spain, Sweden, Switzerland, the UK and West Germany.
7. As per the Convention for the Establishment of a European Space Agency (hereafter ESA Convention), Paris, done May 30, 1975, entered into force October 30, 1980; 1297 UNTS 161; UKTS 1981 No. 30; Cmnd. 8200; 14 ILM 864 (1975); *Space Law – Basic Legal Documents*, C.I.1. See further, e.g. F.G. von der Dunk, "European Space Law", in F.G. von

der Dunk and F. Tronchetti (eds.), No. 3, pp. 208-11; K. Madders, *A New Force at a New Frontier: Europe's Development in the Space Field in the Light of Its Main Actors, Policies, Law and Activities from Its Beginnings up to the Present*, Cambridge University Press, Cambridge, 1997, pp. 39-178; G. Lafferranderie, *European Space Agency*, in J. Wouters (ed.), *International Encyclopaedia of Laws: Intergovernmental Organisations*, Kluwer Law International, Alphen aan den Rijn, 2005, pp. 17-24.

8. See "European Space Agency", *Wikipedia*, at https://en.wikipedia.org/wiki/European_Space_Agency (Accessed July 22, 2016).
9. ESA Convention, No. 7, Art. II; emphasis added.
10. See F.G. von der Dunk, No. 7, pp. 228-32; K. Madders, No. 7, pp. 235-41, 520-6. Arianespace was created by the *Statuts de la Société Arianespace* of March 26, 1980.
11. See F.G. von der Dunk, No. 7, pp. 232-6; K. Madders, No. 7, pp. 252-60, 504-14; EUTELSAT was originally established as a separate international organization by the Convention Establishing the European Telecommunications Satellite Organisation (EUTELSAT), Paris, done July 15, 1982, entered into force September 1, 1985; UKTS 1990 No. 15; Cm. 956; Cmnd. 9069; *Space Law – Basic Legal Documents*, C.II.1. In 1999, its constitutional structure was fundamentally changed so as to turn it into a basically private satellite communication provider Eutelsat overseen by a rudimentary EUTELSAT IGO; see F.G. von der Dunk, "International Organizations in Space Law", in F.G. von der Dunk and F. Tronchetti (eds.), No. 3, pp. 317 ff.
12. See F.G. von der Dunk, No. 7, pp. 236-8; K. Madders, No. 7, pp. 264-70, 516-20. EUMETSAT was established by the Convention for the Establishment of a European Organisation for the Exploitation of Meteorological Satellites (EUMETSAT), Geneva, done May 24, 1983, entered into force June 19, 1986; as amended July 14, 1994, entered into force July 27, 1994; UKTS 1999 No. 32; Cm. 1067; Cmnd. 9483; 44 ZLW 68 (1995).
13. See F.G. von der Dunk, No. 7, pp. 258-62.
14. Ibid., pp. 262-5.
15. As for Galileo, already a GNSS Supervisory Authority (GSA) has been established by means of Council Regulation on the establishment of structures for the management of the European satellite radio-navigation programmes, No. 1321/2004/EC, of July 12, 2004; as for Copernicus (originally labelled Global Monitoring for Environment and Security; GMES), so far it is directly coordinated and managed by the European Commission itself.
16. The ECSC was established by the Treaty establishing the European Coal and Steel Community, Paris, done April 18, 1951, entered into force July 23, 1952, expired July 23, 2002; 126 UNTS 140.
17. The EAEC was established by the Treaty establishing the European Atomic Energy Community, Rome, done March 25, 1957, entered into force January 1, 1958; 298 UNTS 167.
18. The EEC was established by the Treaty of Rome, or Treaty establishing the European Economic Community, Rome, done March 25, 1957, entered into force January 1, 1958; 298 UNTS 11.
19. See F.G. von der Dunk, No. 7, pp. 239-44; R.H. Lauwaars, "Institutional Structure", in P.J.G. Kapteyn et al. (eds.), *The Law of the European Union and the European Communities*, 4th ed., Kluwer Law International, Alphen aan den Rijn, 2008, pp. 180-252; A. Arnull et al., *Wyatt & Dashwood's European Union Law*, 5th ed., Sweet & Maxwell, London, 2006, pp. 32-51, 387 ff.
20. See esp. Treaty on European Union, Maastricht, done February 7, 1992, entered into force November 1, 1993; 31 ILM 247 (1992); OJ C 191/1 (1992).

21. See further F.G. von der Dunk, No. 7, pp. 244-5; K. Madders, No. 7, pp. 566-86; G. Lafferranderie, No. 7, pp. 139-54.
22. See Single European Act, Luxembourg/The Hague, done February 17/28, 1986, entered into force July 1, 1987, Art. 24; UKTS 1988 No. 31; Cm. 372; OJ L 169/1 (1987); 25 ILM 506 (1986).
23. Commission Directive amending Directive 88/301/EEC and Directive 90/388/EEC in particular with regard to satellite communications (hereafter Satellite Directive), 94/46/ EC, of October 13, 1994; OJ L 268/15 (1994). See further, e.g. F.G. von der Dunk, No. 7, pp. 246-9.
24. Cf. Satellite Directive, No. 23, Art. 3.
25. Directive of the European Parliament and of the Council on the legal protection of databases, 96/9/EC, of March 11, 1996; OJ L 77/20 (1996). See further, e.g. F.G. von der Dunk, No. 7, pp. 249-50.
26. See further on this, F.G. von der Dunk, No. 7, pp. 251ff.
27. See L.J. Smith, "Legal Aspects of Satellite Navigation", in F.G. von der Dunk and F. Tronchetti (eds.), No. 3, pp. 561-5, 574-5; also F.G. von der Dunk, No. 7, pp. 258-62; G. Lafferranderie, No. 7, pp. 149-51; K. Madders, No. 7, pp. 547-550.
28. Notably, this concerns Council Regulation setting up the Galileo Joint Undertaking, No. 876/2002/EC, of May 21, 2002; OJ L 138/1 (2002); Council Regulation No. 1321/2004/ EC, No. 15; Regulation of the European Parliament and of the Council on the further implementation of the European satellite navigation programmes (EGNOS and Galileo), No. 683/2008/EC, of July 9, 2008;(OJ L 196/1 (2008); Regulation of the European Parliament and of the Council setting up the European GNSS Agency, repealing Council Regulation (EC) No 1321/2004 on the establishment of structures for the management of the European satellite radio navigation programmes and amending Regulation (EC) No 683/ 2008 of the European Parliament and of the Council, No. 912/2010/EU, of September 22, 2010; OJ L 276/11 (2010); Decision of the European Parliament and of the Council on the rules for access to the public regulated service provided by the global navigation satellite system established under the Galileo programme, No. 1104/2011/EU, of October 25, 2011; OJ L 287/1 (2011); and Regulation of the European Parliament and of the Council on the implementation and exploitation of European satellite navigation systems and repealing Council Regulation (EC) No. 876/2002 and Regulation (EC) No. 683/2008 of the European Parliament and of the Council, No. 1285/2013/EU, of December 11, 2013; OJ L 347/1 (2013).
29. See Regulation No. 1285/2013/EU, No. 28, Art. 15.
30. Ibid., Art. 6.
31. See further F.G. von der Dunk, No. 7, pp. 262-5; G. Lafferranderie, No. 7, pp. 152-3.
32. Regulation of the European Parliament and of the Council on the European Earth monitoring programme (GMES) and its initial operations (2011 to 2013), No. 911/2010/ EU, of September 22, 2010; OJ L 276/1 (2010); Commission Delegated Regulation supplementing Regulation (EU) No 911/2010 of the European Parliament and of the Council on the European Earth monitoring programme (GMES) by establishing registration and licensing conditions for GMES users and defining criteria for restricting access to GMES dedicated data and GMES service information, No. 1159/2013/EU, of July 12, 2013; OJ L 309/1 (2013); and Regulation of the European Parliament and of the Council establishing the Copernicus Programme and repealing Regulation (EU) No 911/2010, No. 377/2014/ EU, of April 3, 2014; OJ L 122/44 (2014).
33. See Regulation No. 377/2014/EU, No. 32, Art. 10.

34. As per Art. 28, Regulation No. 377/2014/EU (*supra*, No. 32), the Union again will own all assets associated with the Copernicus programme, and as per Art. 9 will maintain general control over the process of developing, deploying and operating the space infrastructure.
35. Cf. e.g. F.G. von der Dunk, "Towards One Captain on the European Spaceship – Why the EU Should Join ESA", *Space Policy*,19, 2003, pp. 83-6.
36. See further on this, F.G. von der Dunk, No. 7, pp. 251-5.
37. Framework Agreement between the European Community and the European Space Agency, Brussels, done November 25, 2003, entered into force May 28, 2004; OJ L 261/64 (2004); *Zeitschrift für Luft- und Weltraumrecht*, 53, 2004, p. 89.
38. See further on this, Von der Dunk, No. 7, pp. 255-8.
39. Treaty of Lisbon amending the Treaty on European Union and the Treaty establishing the European Community (hereafter Treaty of Lisbon), Lisbon, done December 13, 2007, entered into force December 1, 2009; OJ C 306/1 (2007).
40. Treaty establishing the European Community as amended by the Treaty of Lisbon amending the Treaty on European Union and the Treaty establishing the European Community, Lisbon, done December 13, 2007, entered into force December 1, 2009, Art. 189; OJ C 326/47 (2012); emphasis added.
41. S. Hobeet al., "A New Chapter for Europe in Space", *Zeitschrift für Luft- und Weltraumrecht*, 54, 2005, p. 347.
42. Act on Space Activities, 1982: 963, November 18, 1982; *National Space Legislation of the World*, Vol. I (2001), at 398; *Space Law – Basic Legal Documents*, E.II.1; *Zeitschrift für Luft- und Weltraumrecht*, 36, 1987, p. 11.
43. Outer Space Act, July 18, 1986, 1986 Chapter 38; *National Space Legislation of the World*, Vol. I (2001), at 293; *Space Law – Basic Legal Documents*, E.I; *Zeitschrift für Luft- und Weltraumrecht*, 36, 1987, p. 12.
44. Law on the Activities of Launching, Flight Operations or Guidance of Space Objects, September 17, 2005, adopted June 28, 2005; *Nationales Weltraumrecht / National Space Law* (2008), at 183.
45. Law Incorporating Rules Concerning Space Activities and the Establishment of a Registry of Space Objects, January 24, 2007; 80 *Staatsblad* (2007), at 1; *Nationales Weltraumrecht / National Space Law* (2008), at 201.
46. Law on Space Operations (*Loi relative aux opérations spatiales*); *Loi n° 2008-518 du 3 juin 2008*; unofficial English version, *Journal of Space Law*, 34 (2), Winter 2008, p. 453.
47. Austrian Federal Law on the Authorisation of Space Activities and the Establishment of a National Space Registry (*Bundesgesetz über die Genehmigung von Weltraumaktivitäten und die Einrichtung eines Weltraumregisters* (*Weltraumgesetz*)), as adopted by Parliament on December 6, 2011; Federal Law Gazette of December 27, 2011; *Zeitschrift für Luft- und Weltraumrecht*, 61, 2012, pp. 37-42, 56 61.
48. Law on activities in outer space (*Lov om aktiviteter i det ydre rum*), Passed by Parliament with the third treatment, May 3, 2016; Parliament Gazette, 2015-17, No. L 128.
49. See F.G. von der Dunk, "The EU Space Competence as per the Treaty of Lisbon: Sea Change or Empty Shell?", *Proceedings of the International Institute of Space Law 2011*,2012, pp. 382-92.
50. See Treaty on European Union as amended by the Treaty of Lisbon amending the Treaty on European Union and the Treaty establishing the European Community, Lisbon, done December 13, 2007, entered into force December 1, 2009, Art. 2; OJ C 326/13 (2012).
51. See Outer Space Treaty, No. 2, Arts. I, II, III, IX, X, XI.
52. Ibid., Art. IV.

53. Ibid., Art. V.
54. Ibid., Arts. VI, VII.
55. Ibid., Art. IX.
56. Ibid., Art. XII.
57. Ibid., Art. III.
58. This very elaborate and detailed regime stems from the three main constitutive documents of the ITU: the Constitution of the International Telecommunication Union, Geneva, done December 22, 1992, entered into force July 1, 1994; 1825 UNTS 1; UKTS 1996 No. 24; Cm. 2539; ATS 1994 No. 28; Final Acts of the Additional Plenipotentiary Conference, Geneva, 1992 (1993), at 1; the Convention of the International Telecommunication Union, Geneva, done December 22, 1992, entered into force 1 July 1994; 1825 UNTS 1; UKTS 1996 No. 24; Cm. 2539; ATS 1994 No. 28; Final Acts of the Additional Plenipotentiary Conference, Geneva, 1992 (1993), at 71; and the Radio Regulations, at www.itu.int/pub/R-REG-RR-2012 (Accessed July 21, 2016).
59. See in general M. Gerhard, "Article VI", in S. Hobe, B. Schmidt-Tedd and K.U. Schrogl (eds.), *Cologne Commentary on Space Law*, Vol. I, Carl Heymanns Verlag, Cologne, 2009, pp. 111-22.
60. Outer Space Treaty, No. 2, Art. XIII.
61. Outer Space Treaty, No. 2, Art. VI.
62. See e.g. P. Jankowitsch, No. 3, pp. 5-6.
63. Convention on International Liability for Damage Caused by Space Objects, London/Moscow/Washington, done March 29, 1972, entered into force September 1, 1972; 961 UNTS 187; TIAS 7762; 24 UST 2389; UKTS 1974 No. 16; Cmnd. 5068; ATS 1975 No. 5; 10 ILM 965 (1971).
64. Convention on Registration of Objects Launched into Outer Space, New York, done January 14, 1975, entered into force September 15, 1976; 1023 UNTS 15; TIAS 8480; 28 UST 695; UKTS 1978 No. 70; Cmnd. 6256; ATS 1986 No. 5; 14 ILM 43 (1975).

7

50th Anniversary of the Outer Space Treaty: US Contributions as We Move into the 21st Century

Philip A. Meek

> "*More than any other imaginative concept, the mind of man is aroused by the thought of exploring the mysteries of outer space. Through such exploration, man hopes to broaden his horizons, add to his knowledge, improve his way of living on earth.*"
>
> —Dwight Eisenhower, June 20, 1958[1]

Introduction

Fifty years is an extremely short time span in the annals of history. And yet, over the past half century of operations and activities in outer space since the inception of the Outer Space Treaty (OST)[2], any casual observer will stand in amazement at the significant achievements that mankind has accomplished in the space arena. The early days of the space age were characterised by an intense space race between the two superpowers of the Cold War as each strived to be the first to launch a satellite into orbit, the first to orbit a man around the earth, the first to walk in space and the first to set foot on the Moon.

At its core, the OST is primarily concerned with the exploration and exploitation of space, of the celestial bodies of our solar system and beyond.

The US has a rich history of human space exploration based upon the principles of the OST. The following discussion will highlight selected US contributions under the OST, and serve as a springboard for discussions about two recent US initiatives that push the envelope as far as certain interpretations of the Treaty are concerned.

Human Exploration of Outer Space: The Race to the Moon

Project Mercury (1958-1963) was America's first human spaceflight programme undertaken by the National Aeronautics and Space Administration (NASA).[3] The objectives of the Mercury programme were to orbit a manned spacecraft around the world, investigate man's ability to function in space, and to recover both man and the spacecraft safely. Astronaut Alan B. Shepard, Jr., became the first American to fly in space on May 5, 1961, aboard the Freedom 7 Mercury capsule. In 1962, Astronaut John Glenn completed the first US manned space flight in low earth orbit on a Mercury mission.

Three weeks after Astronaut Shepard's historic space flight, President John F. Kennedy surprised the world by announcing that the US was embarking on a goal to send astronauts to land on the Moon and return safely to Earth before the end of the decade. This was an ambitious goal that required significant technical breakthroughs and the development of new crew skill sets. The Mercury programme also set in motion our continuing quest to land an astronaut on Mars and travel further into deep space.

Project Gemini (1965-66)[4] fulfilled part of that goal by developing missions that successfully completed additional building blocks for the trip to the Moon, including longer flight durations in space and understanding the resulting effects on the astronauts, and learning to rendezvous and dock two spacecraft in orbit. Gemini IV produced America's first spacewalk by Astronaut Ed White. In addition, five Lunar Orbiter missions mapped the surface of the Moon in preparation for the arrival of astronauts several years later.

The primary goals of NASA's third human spaceflight programme, Project Apollo (1969-72)[5] were to land a man on the Moon, carry out a program of scientific exploration of the Moon, develop man's capability to work in the lunar environment, and return the crew safely to Earth.

On the Apollo 11 mission, Astronaut Commander Neal Armstrong became the first man to set foot on the Moon on July 20, 1969, a spectacular accomplishment viewed by the world in real-time on television. Astronauts

Armstrong and Michael Collins explored the Moon's surface, collected rocks and other geologic materials to take back to Earth and set up lunar experiments that would continue for decades to come to learn about the lunar environment. Several days later, they blasted off the Moon's surface and returned safely to Earth, thus fulfilling the lofty goal articulated by President Kennedy eight years previously. Subsequent Apollo missions to the Moon were devoted to lunar research. Astronaut Gene Cernan, on the final Apollo mission in 1972, was the last man to set foot on the Moon.

Skylab – America's First Space Station

Skylab 1, America's first space station, was launched into orbit in May 1973, where it remained operational until 1979.[6] The primary goals of the Skylab Programme were to take the knowledge obtained from earlier space programmes and conduct follow-on studies in microgravity, make Earth observations and expand the knowledge of solar astronomy. The most important goal, however, was to prove that humans could exist in space for the extended periods of time necessary to travel into deep space to Mars and other celestial bodies. The successful Skylab Programme represented the transition between the Apollo Programme and the Space Shuttle.

The Apollo Soyuz Test Project

America's manned Skylab was complemented in 1975 by the Apollo Soyuz Test Project, the world's first internationally crewed (American and Soviet) space mission.[7] It was a remarkable geopolitical achievement to see the two adversarial super powers of the Cold War cooperating peacefully during a period of détente to participate in the Apollo Soyuz joint space project, placing their astronauts and cosmonauts in the same cramped habitat in outer space. This mission basically ended the space race between the US and USSR that had existed since 1957, when the Soviets launched Sputnik.[8] Significantly, these two living habitats in space served as the forerunners of the US International Space Station.

Space Transportation System: The Space Shuttle

In April 1981, the US launched its first space shuttle, Columbia. The space shuttle, officially called the Space Transportation System, was important to US space exploration efforts because it was the first reusable launch platform. Over its 30-year lifetime, the space shuttle, comprising five spacecraft – Columbia, Challenger, Discovery, Atlantis and Endeavour – flew over 135 missions, carried 850 people (355 individuals) into orbit; launched, recovered and repaired satellites; conducted state of the art space research; and built the International Space Station (ISS) in orbit, the largest structure in space.[9]

The final shuttle flight was completed in 2011. For the past five years, the US has relied on Russian Soyuz spacecraft to reach the ISS until a follow-on shuttle is built. NASA's replacements for the space shuttle will be Boeing's Crew Space Transportation (CST-100) and SpaceX's Dragon, the first private spacecraft to carry crew members to the ISS commencing in 2017 if all goes as planned.[10]

Living in Space: The ISS

The construction of the ISS in low Earth orbit commenced in November 1998 with the launch of the Zarya control module atop a Proton Rocket.[11] Building the ISS required 36 space shuttle assembly flights and five Russian launches.[12] The first operational crew, Expedition 1, arrived at the ISS in November 2000 on Soyuz TM-3. Construction on the ISS was completed in 2011.

The ISS comprises five participating space agencies, namely NASA, Russia's Roscosmos State Corporation for Space Activities (Roscosmos), Japan's Aerospace Exploration Agency Space Agency (JAXA), the European Space Agency (ESA) and the Canadian Space Agency (CSA). These five space agencies represent 15 countries that built the ISS at a cost of $100 billion and continue to operate it today.[13]

"Since November 2000, more than 220 people from 17 countries have visited the ISS, and the orbiting laboratory has hosted more than 1,700 research investigations from more than 80 countries. A total of 16 people lived and worked aboard the ISS in 2015, conducting hundreds of scientific investigations. This is a critical step in our journey to Mars."[14]

Current plans are for the ISS to remain operational through at least 2020; however, NASA has requested an extension until 2024. Because of the success in conducting the wide variety of research projects at the ISS which are conducted in a micro-gravity environment that cannot be replicated on Earth, some nations would like to see the ISS remain operational well into the 2030s to conduct additional scientific experiments, and to gain more experience in long-term living in space.[15]

Deep Space Exploration

NASA plans to send humans to an asteroid by the mid-2020s, and to Mars in the 2030s.[16] Of note, US Astronaut Scott Kelly and Roscosmos' Mikhail Kornienco recently completed living aboard the ISS for a full year during 2015-2016. The knowledge gained concerning human physiology changes resulting from their continuous presence in space for such an extended period of time will aid immeasurably in preparing future space travellers for the much longer journeys when venturing into deep space to visit asteroids and for the trip to Mars.

To reach Mars, asteroids and other celestial bodies, NASA is developing a new Space Launch System (SLS) with the most powerful rocket motors ever developed for the long trips into deep space. NASA is also developing a new spacecraft, Orion, to serve as the exploration vehicle that will carry the crew to space, provide emergency abort capability, sustain the crew during space travel, and provide safe return from deep space return velocities.[17] Orion has already completed several initial flight tests.

NASA has been studying the Mars environment through flybys, probes and robotic landers such as the Opportunity and Curiosity rovers, and the Mars Reconnaissance Orbiter for over 40 years in preparation for the flight to Mars in the future. NASA has also conducted or is conducting many other deep space missions to explore our solar system, including missions to the Sun, Mercury, Venus, Jupiter, Saturn, Uranus, Neptune, Pluto, the dwarf planet Ceres and several asteroids.

US Commercial Space Launch Competitiveness Act

In November 2015, President Obama signed into law the US Commercial Space Launch Competitiveness Act (CSLCA).[18] The CSLCA gives US individuals and companies, and foreign companies if the controlling interest is held by a US citizen or entity, the right to conduct a wide range of commercial space activities, including but not limited to: mining asteroids and other celestial bodies for precious metals, minerals and water; selling the resources that are mined; and keeping the financial rewards from the mining endeavour. Thus, the statute opens the door to foreign investment by business entities in non-space power countries to access space and benefit from this new space industry.

The statute specifically provides that these activities "will be conducted in accordance with the international obligations of the United States and subject to the authorisation and continuing supervision by the Federal Government". Further, the statute clearly states, "It is the sense of Congress that by the enactment of this Act, the United States does not thereby assert sovereignty or sovereign or exclusive right or jurisdiction over, or the ownership of, any celestial body." These provisions are entirely consistent with the US' obligations under the OST and the treaty itself.

This bold initiative by the US creating domestic law that authorises commercial activities in outer space should not come as a surprise to the international community. The US consistently articulated its position that such activities were permitted during the negotiations in the United Nations Committee on the Peaceful Uses of Outer Space (UNCOPUOS) in the 1960s for the OST, and subsequently during the negotiations for the Moon Agreement in the 1970s. The US has not wavered from that position in subsequent years.

Any new paradigm such as the CSLCA is bound to be met with resistance in some quarters if for no other reason than it represents a change in the comfortable routine of private, commercial and government activities in the realm of outer space. Predictably, some in the international space community raised academic, legal and policy objections to the US statute. Delving into the range of objections and concerns voiced as to the legality of the US statute is beyond the scope of this paper. Suffice it to say that most of the arguments centre on interpretations of Article I and Article II of the OST.

Basically, the legal issues centre around: (1) the meaning of the word "use" in Articles I and II; (2) the meaning of the term "province of all mankind" in Article 1, and (3) whether the mining and removal of resources constitute an "appropriation" of the asteroid or other celestial body, which is prohibited pursuant to Article II of the OST.

Although countless articles on resource mining in outer space have been written in the news media, law school journals and space-related internet blogs, and further that the issue of outer space mining has been the subject of space conferences for decades, substantive multi-lateral discussions at the UN or national representative levels have not borne fruit to resolve any of the definitional issues or practical problems raised pursuant to the language of the OST.[19]

Into this legal vacuum stepped the US with its new statute. Space mining is no longer a theoretical idea as it was in 1967 when the OST came into effect. Technology has progressed to the point where space mining could occur within the next decade. While the statute may be criticised as a unilateral action taken by the US Government, rather than a multilateral resolution of the meaning of the applicable terms in the OST, the US action could be quite helpful as a "forcing factor" in moving these issues to the forefront at the international level to resolve some of the most important issues concerning 21st Century exploration and exploitation of space. Further, the statute may provide a more practical and expedient way to move ahead in the relatively near future gaining experience with space mining activities, rather than getting bogged down for years in unproductive formal negotiations that rehash the contentious and problematic "common heritage of mankind" philosophy of the Moon Agreement.

The statute could stimulate private company, commercial, and bilateral/multilateral discussions with the US and within UNCOPUOS as the commercial mining technology and space mining activities approach initial operational capability, and as the US grants mining licences to US commercial entities or approves foreign participation in a US entity. Joint ventures between foreign entities around the globe and US companies would enable the US companies to spread their financial risk, and the foreign ventures could participate in the fruits of the mining effort without absorbing the full development and operational costs, a win-win situation for all parties. This represents a significant opportunity for non-space powers and their citizens to partner with US private industry in new enterprises in outer space, to develop their own niche space industries, and thereby create job and revenue generators for their national economies.

Another benefit of the US statute would be that the "real world" space mining operations and activities may develop technologies, techniques, procedures and methods of mining celestial bodies that become accepted customary industry practices, which in turn could create a more informed basis for negotiating binding treaty provisions in the future.

The Protection of Cultural Heritage in Space

On July 20, 2011, NASA released its "Recommendations to Space-Faring Entities: How to Protect and Preserve the Historic and Scientific Value of U.S. Government Lunar Artifacts". Although no human has set foot on the Moon since the final US Apollo mission in 1972, NASA was rightly concerned that numerous commercial entities and States were planning lunar missions, including manned and unmanned landings on the Moon, robotic travel on the surface of the Moon, research and scientific experiments.

For instance, the Google Lunar X Prize will award $30 million total in prizes, with the first place going to a privately-funded team that builds a rover that successfully lands on the Moon, explores it by moving at least one third of a mile and returns high definition video and imagery back to Earth.[20] Further, both the Russians and the Chinese are planning manned lunar missions within the next few years.

While the US still owns all of the physical objects left on the Moon by their astronauts pursuant to provisions in the OST, there are uncertainties as to how the non-physical scientific, cultural and historic aspects of their visit to the lunar surface, such as mankind's first footprints on the Moon, lunar rover tracks and the seven US Apollo landing sites at Tranquility Base, can be protected against deterioration and destruction caused by new visitors, whether robotic or human, to the Moon's surface. How was NASA to protect the US space heritage, and in many ways the world space heritage, on the Moon?

Because no international agreements exist concerning the preservation of historic and cultural items and sites in outer space, NASA sought advice from in-house and external experts, including foreign space agencies and commercial entities. After considering their input, NASA issued the Recommendations which, by the terms of the document, are applicable only to US Government lunar artefacts. The Recommendations specifically recognise that:

> Until more formal USG guidance is developed and perhaps a multilateral approach is developed to reflect various nations' views on lunar hardware of scientific and historic value, NASA has assembled this document that contains the collected technical knowledge of its personnel – with advice from external experts and potential space-faring entities – and provides interim recommendations for lunar vehicle design and mission planning teams. As such, this document does not represent mandatory USG or international requirements; rather, it is offered to inform lunar spacecraft mission planners in helping preserve lunar historic artefacts and potential science opportunities for future missions.[21]

The NASA Recommendations also include a Legal Framework section which concludes that the Recommendations are consistent with international law, and then lists various provisions of the OST as authority.[22] However, a supporting legal analysis is not offered. The failure to include a convincing legal analysis is disappointing since these NASA Guidelines are the initial guidelines of any nation that has sought to protect its objects and sites on the lunar surface. This significant document of first impression represents a golden opportunity for the US to explain to the global community, especially those nations or academics that may question the US initiative, how these

Recommendations are consistent with the OST, rather than leave it to the reader to discern. This is particularly so in the case of protection of geographic lunar sites, rather than US physical objects.

Nevertheless, considering the "soft" and temporal non-binding language in the recommendations quoted above, there is little to criticise in the US initiative. It is a conservative document that recognises the sensitivities surrounding this first initiative of its kind in space, yet provides a modicum of protection for US artefacts and exploration sites on the lunar surface. At a minimum, the Recommendations could serve as a template to jump-start serious discussions within the international community on protections for our outer space heritage.

Conclusion

The US human exploration of space has proceeded from our first astronaut in space, to the first landing of a human on the lunar surface, to living in space for over a year, all in the short span of 58 years. During this time, the US has always been guided by and respected the principles of the OST and will continue to do so in the future.

Within the next decades, humans will land on Mars and asteroids in deep space. Scientific probes and robotic landers to distant planets, their moons and other celestial bodies have already been sent to the limits of our solar system, with missions planned for many years into the future.

Our initiatives through the US Commercial Space Launch Competitiveness Act will create new opportunities for non-spacefaring States and the global private sector to participate in the peaceful exploration and exploitation of outer space through avenues not previously available. This is an exciting time for space exploration!

The passing of the Cold War Era and the transition through subsequent global challenges have not minimised the significance of the OST and its progeny – the Rescue and Return Agreement, the Registration Agreement and the Liability Agreement – to the world community. Rather, this bedrockspace treaty remains as vibrant and relevant today as it was at its inception 50 years ago. We celebrate the 50th Anniversary of the OST and look ahead to the challenges we will encounter in space for the next 50 years.

ENDNOTES

1. He was the 34th President of the USA (1953-61). Quoted in the Introduction of the National Space Policy of the United States of America, June 28, 2010, p. 1.
2. Treaty on Principles Governing the Activities of States in the Exploration and Use of Outer Space, including the Moon and Other Celestial Bodies, January 27, 1967, 18 U.S.T. 2410, 610 U.N.T.S. 205.
3. "About Project Mercury", National Aeronautics and Space Administration (NASA), at www.nasa.gov/mission_pages/mercury/missions/program-toc.html (Accessed July 8, 2016).
4. "Project Gemini", NASA, at www.nasa.gov/mission_pages/gemini/missions/index.html (Accessed July 17, 2016). The name Gemini, Latin for "twins", recognised that the new crew capsule carried two astronauts, rather than the single astronaut aboard the Mercury capsule.
5. "The Apollo Missions", NASA, at www.nasa.gov/mission_pages/apollo/missions/index.html (Accessed July 17, 2016).
6. "Skylab: America's First Space Station", NASA, at www.nasa.gov/mission_pages/skylab/missions/index.html (Accessed September 1, 2016).
7. "A Brief History of Space Exploration: Into Orbit", Aerospace, at www.aerospace.org/education/stem-outreach/space-primer/a-brief-history-of-space-exploration, (Accessed July 8, 2016).
8. Richard J. Samuels (ed.), *Encyclopedia of United States National Security*, 1st ed., Sage Publications, 2005, p. 669. "Most observers felt that the U.S. moon landing ended the space race with a decisive U.S. victory.... The formal end of the space race occurred with the 1975 joint Apollo-Soyuz mission, in which U.S. and Soviet spacecraft docked, or joined, in orbit while their crews visited one another's craft and performed joint scientific experiments." Cited in "Apollo-Soyuz Test Project", *Wikipedia*, Reference 1.
9. "NASA Space Shuttle Era Facts", NASA, at www.nasa.gov/pdf/566250main_SHUTTLE (Accessed September 6, 2016).
10. Mark Prigg, "Blast Off! NASA Reveals Boeing and SpaceX Will BOTH Build Shuttle Replacement in $6.8 Billion Deal for America's 'Space Taxi'", MailOnline, at www.dailymail.co.uk/sciencetech/article-2758436/were-space-nasa-reveals-Boeing-SpaceX-build-americas-new-taxi-shuttle-finally replaced (Accessed September 6, 2016).
11. "Zarya", NASA, at www.nasa.gov/mission_pages/station/structure/elements/fgb.html (Accessed September 6, 2016).
12. *Reference Guide to the International Space Station*, Utilisation Edition, NASA, September 2015, at www.nasa.gov/sites/default/files/atoms/np-2015-05-022-ISS-guide-2015-update-111015-508c-pdf, p. 78 (Accessed September 6, 2016).
13. Tim Sharp, "International Space Station: Facts, History and Tracking", Space.com, April 5, 2016, at www.space.com/spaceflight (Accessed August 23, 2016).
14. Statement by Brian Israel, United States Representative, on Agenda Item 3, "General Exchange of Views", before the United Nations Committee on the Peaceful Uses of Outer Space, April 4, 2016, p. 2.
15. Tim Sharp, No. 13.
16. "NASA's Journey to Mars", NASA, December 1, 2014, at www.nasa.gov/content/nasas-journey-to-mars (Accessed September 6, 2016).
17. "Orion Overview", NASA, at www.nasa.gov/exploration/systems/orion/about/index.html (Accessed September 9, 2016).
18. Public Law 114-90, November 25, 2015.

19. The Agreement Concerning the Activities of States on the Moon and Other Celestial Bodies (hereinafter the Moon Agreement), December 5, 1979, 1363 U.N.T.S. 3, did address several of these issues through its "common heritage of mankind" language; however, the problematic Agreement has been ratified by only 16 States, none of which is a space power. The US has neither signed nor ratified the Moon Agreement and is not bound by its provisions. All factors considered, it is a failed treaty that has been soundly rejected by the vast majority of the international community. For a very good discussion comparing the "province of all mankind" phrase to the "common heritage of mankind", see J.I. Gabrynowicz, "The 'Province' and 'Heritage' of Mankind Reconsidered: A New Beginning", in W.W. Mendell (ed.), *Second Conference on Lunar Bases and Space Activities of the 21st Century, Proceedings from a Conference held in Houston, Texas, April 5-7, 1988*, NASA Conference Publication, 1992, p. 691.
20. "NASA Offers Guidelines to Protect Historic Sites on the Moon", NASA Press Release 12-168, May 24, 2012, at www.nasa.gov/home/hqnews/2012/may/HQ_12-168_Lunar_Protection_ Guidelines (Accessed September 1, 2016).
21. "NASA's Recommendations to Space-Faring Entities: How to Protect and Preserve the Historic and Scientific Value of U.S. Government Lunar Artifacts", NASA, July 20, 2011, Section A1-Preface, Authority, and Definitions, p. 5.
22. Ibid., Legal Framework, p. 6.

8

Russia and the Outer Space Treaty

Aleksandr Klapovskiy and Vladimir Yermakov

The 1967 Treaty on Principles Governing the Activities of States in the Exploration and Use of Outer Space, including the Moon and Other Celestial Bodies (in short, the Outer Space Treaty, or OST) holds a special place in the theory and practice of legal regulation of international space activities.

Going in for an assessment of the OST historical and actual practical importance one should keep in mind that it was the first international agreement to mark the start of a space era in the evolution of humanity. During that time all states and all peoples on Earth did face a totally new strategic task – a systematic study, exploration and use of outer space in the interests of peaceful development and prosperity of mankind. National space programmes, initiated in many countries, as well as programmes of international space cooperation demonstrated impressive opportunities of using outer space for scientific research, social economic development and protection of environment. Meanwhile, it was impossible to enter into the privileges of enjoying the benefits of outer space exploitation without taking additional responsibilities. The prospects for space exploration brought about brand new problems to be solved – a necessity for the prevention of an arms race in outer space (PAROS) by the prevention of the placement of weapons in outer space (PPW).

Actually, from the very first steps of space exploration a new track of space activities related to national defence started to emerge. The US and USSR, and later a number of other states, conducted extensive research into a possibility of space technology application for military purposes. A wide range of space military systems, known under the common name of support space systems, were built and put into service.

A possibility of placement of weapons in outer space to be aimed at land, sea or air based targets was thoroughly explored. The research conducted by a number of countries at that time did not prove any superiority of possible space weapon systems over traditional means of warfare, and, therefore, they did not receive further visible development. Nevertheless, some types of missile weapon systems capable of orbiting the Earth were built and did exist for a while.

Besides, much attention was paid to the exploration of an opportunity of building space-based missile defence systems. Some achievements in that field were marked in the framework of the US Strategic Defence Initiative, but that did not allow building specific space-based missile defence systems because of tremendous costs and huge technological shortcomings. Extensive research was also conducted in the field of anti-satellite (ASAT) systems. The USSR and US carried out some ASAT projects.

In general, it was at the very initial stage of outer space exploration that the international community became aware of the need for a legal regulation. The OST became a product of different unilateral, bilateral and multilateral initiatives, intensive discussions and negotiations. At the same time, a number of other agreements were prepared. Afterwards, the OST served as a basis for all other treaties, agreements and conventions in the field of outer space activities. Some of them did develop the provisions of the 1967 Treaty, others regulated the related issues. Thus, all of them formed the contemporary International Space Law (ISL).

The USSR and US were the first to embark on a full-scale outer space exploration and, simultaneously, they started a close control over such activity The principled positions and ambitious initiatives of the two leading spacefaring powers played a crucial role in the ISL development. In general, the Soviet and American initiatives at the dawn of the space age reflected the depth of their understanding of the issues related to space exploration, as well as the burden of responsibility taken in this respect, including that in the PAROS/PPW context.

For example, back in the 1950s, the USSR came up with a proposal to fully exclude outer space from military use with a simultaneous dismantlement of all US military bases, located outside the US national territory, and go together in for a complete nuclear disarmament. In 1957, at the United

Nations (UN) Disarmament Commission, the USSR proposed to adopt an international convention on the reduction of armaments and armed forces, banning nuclear and thermonuclear weapons, non-use of missiles capable of carrying nuclear and thermonuclear warheads and banning the use of guided missiles for military purposes. In 1958, the USSR submitted to the UN a draft memorandum prohibiting the use of outer space for military purposes with a simultaneous dismantlement of all foreign military bases on the territory of other states and development of international cooperation in the field of outer space exploration. In 1959, the USSR introduced in the UN a proposal to end production of and destroy all types of missile weapons of any range, including space missiles of military nature, and eliminate the appropriate missile launchers. In 1962, the USSR introduced in the UN a draft treaty on general and complete disarmament under the international control. The draft included such measures as a gradual destruction of intercontinental ballistic missiles, missile defence systems, dismantlement of launching platforms, means of missile launching control, destruction of storage depots for such missiles, a ban on launching and placing in outer space systems capable of carrying weapons of mass destruction (WMD).

In 1958, the US finally seemed to respond to the Soviet proposals. The US President in a letter to the Soviet Government advised about a possibility of an American consent to ban the use of outer space for tests of missiles for military purposes, and the production of weapons designed to be used in space. However, it turned out to be not more than a hint that Washington could agree only on some selected elements of the USSR comprehensive proposals, putting all the rest of them in brackets.

Under those conditions, the ISL development went on under the aegis of the UN through elaboration of international norms regulating only certain aspects of space activities. Nevertheless, together with the OST they did form a legal regime of exploration and use of outer space, including for military purposes.

Specific aspects of the use of outer space are now regulated by a number of international legally binding agreements:

- Treaty Banning Nuclear Weapon Tests in the Atmosphere, in Outer Space and under Water (1963);
- Agreement on the Rescue of Astronauts, the Return of Astronauts and the Return of Objects Launched into Outer Space (1968);
- Convention on International Liability for Damage Caused by Space Objects (1972);

- Convention on Registration of Objects Launched into Outer Space (1976);
- Convention on the Prohibition of Military or Any Other Hostile Use of Environmental Modification Techniques (1977); and
- Agreement Governing the Activities of States on the Moon and Other Celestial Bodies (1948).

It should be taken into account that the UN Charter, as a fundamental document in the field of contemporary international law, contains important provisions well applicable to outer space activities.

Besides, a crucially important total ban on space-based missile defence was incorporated into the 1972 USSR-US Treaty on the Limitations of the Anti-Ballistic Missile Defence Systems (the ABM Treaty). However, the US unilaterally withdrew from the ABM Treaty in June 2002.

Nevertheless, it could be stated that decades-long international legal activities finally led to a positive outcome. The development of a wide range of legally binding norms in the field of exploration and use of outer space was carried out within the UN framework.

Summing up all the existing norms and different obligations with regard to space activities and space vehicles, we can point out the following core legal provisions:

- Outer space is open for a free and equal exploration and use by all states without any discrimination.
- Outer space is not subject to national appropriation by claim of sovereignty, by means of use or occupation, or by any other means.
- States agreed to carry out activities in the exploration and use of outer space in accordance with international law, including the UN Charter. The principle of no use of force or threat of force in international relations (Article 1, paragraph 2 of the UN Charter) extends to space activities of the states and ensuing relations;
- States undertook to prohibit, prevent and refrain from nuclear weapon detonations in outer space.
- States undertook not to place any objects carrying nuclear weapons or any other kinds of WMD in orbit around the Earth, install such weapons on celestial bodies or station such weapons in outer space in any other manner. On celestial bodies any establishment of military bases, installations and fortifications, testing of any type of weapons and the conduct of military manoeuvres is totally prohibited.

- States undertook to prohibit military or any other hostile use of means affecting environment, including outer space.
- A state, on whose registry an object launched into outer space is carried, is supposed to retain jurisdiction and control over such object, while in outer space (ownership of objects launched into outer space is not affected by their presence in outer space).
- A state that launches or procures the launching of an object into outer space and each state from whose territory or facility an object is launched is internationally liable for damage to another state.
- If a state is not in a position to identify a space object which has caused damage to it or which may be of a hazardous or deleterious nature, other states possessing space monitoring and tracking facilities are expected to respond to a request by that state for assistance in the identification of the object.
- States undertook not to interfere with the use of national technical means of control of other States' weapons (including the appropriate space means of verification).

Thus, currently there are serious legal limitations on the military use of outer space. There also exist a number of international obligations which protect space assets from hostile interference. All these limitations, however, do not impede (all space powers did point it out) building, deploying and using military support space infrastructure. Such infrastructure, by nature, is not considered to be "destabilising armaments" and, therefore, is believed to be not impeding to strategic balance of forces.

However, one should not forget that the OST fully bans only WMD in outer space. Regrettably, many "loopholes" in the ISL regarding other types of weapons remain unaddressed. Therefore, conceptually, we need to bear in mind that, from a legal point of view, outer space remains unprotected from attempts of turning it into a new arena of military confrontation. Theoretically, non-WMD weapons may be placed into outer space any moment and become weapons of an actual prompt and clandestine use with a global outreach. That could undermine strategic stability and, therefore, upset international security environment. Therefore, urgent additional efforts are needed to develop international legal norms, which could restrict or prohibit a number of potentially dangerous tracks of military technology development related to outer space activities:

- Building and testing (outside outer space) of objects capable of carrying nuclear weapons or other types of WMD, which could be designed for outer space;

- Building, testing and deploying in outer space of conventional weapons (e.g. systems discriminately engaging air-, sea- and land-based targets);
- Building, testing and deploying of ASATs in outer space (for now, there is no prohibition on building, testing, and deploying land-, sea- and air-based ASATs);
- Building, testing and deploying in space of means of jamming of radio-electronic and electro-optic systems of different types of placement, as well as building, testing and deploying of similar means with other kinds of placement (other than space-based) and designed to affect space assets.

The issue of the US global missile defence system unilateral and unlimited deployment demands a special attention in this regard. After the US withdrew from the ABM Treaty, Washington also dropped the obligation not to build, test and deploy space-based missile defence systems. This paved the way to the development of space weapons designed for missile defence, and, given the applied technology similarity, it also provided for an opportunity to conduct parallel work in the field of space-based ASAT.

The loopholes in the legal framework for military space activities have been constantly inspiring additional efforts to develop appropriate international agreements. Without any attempt to give here a comprehensive overview of the political and diplomatic steps taken by the USSR to prevent placement of weapons in outer space in order not to turn outer space into a new military confrontation arena, it is worthwhile to point out three most important initiatives:

- A draft treaty on the prohibition of the stationing of weapons of any kind in outer space (submitted to the UN General Assembly [UNGA] in 1981)
- A draft treaty on the prohibition of the use of force in outer space and from space against the Earth (submitted to the UNGA in 1983 and drafted in 1984 to include the prohibition of the use of force from the Earth against space objects).
- An initiative/political commitment (1983) of the USSR not to be the first to place ASAT weapons (including ASAT test launchers) in outer space (it was a unilateral Moratorium, which would stay in force provided other states refrain from placing ASAT weapons in outer space).

The USSR initiatives related to the prohibition of the use of force in space, from space and against space objects have laid the foundation needed to comprehensively address the problem of weaponisation of outer space and preventing it from becoming a new arena for an armed confrontation.

Specifically, the USSR draft agreements provided for:

- Not to test or deploy by placing in orbit around the Earth, stationing on celestial bodies or in any other manner of any space-based weapons designed to destroy objects on Earth, in atmosphere or in outer space;
- Not to utilise space objects in orbit around the Earth, on celestial bodies or stationed in outer space in any other manner, as means to destroy any targets on Earth, in atmosphere or in outer space;
- Not to destroy, damage, disturb the normal functioning or change the flight trajectory of space objects of other States;
- Not to test or build new ASAT weapons and to destroy all existing systems of that kind;
- Not to test or use any manned spacecraft for military, including ASAT, purposes.

All the initiatives by the USSR, and later Russia, as its continuator, even today have not lost their fundamental value in preventing weaponisation of outer space. Yet, the US never supported any of them. Moreover, in 1985, using its ASAT assets, the US intercepted its own space object (Solwind artificial satellite). Facing such a reality, the USSR had to declare that the 1983 Soviet unilateral moratorium was thereby undermined. Nonetheless, exercising goodwill, the USSR de facto continued to refrain from launching ASAT weapons into outer space. In 1992, the President of the post-Soviet Russia reaffirmed the willingness to eliminate, on mutual basis with the US, all existing ASAT weapons and to work out an agreement on a comprehensive prohibition of weapons, specially designed to destroy satellites. However, even this Russian initiative did not find any positive response from Washington.

Due to the difficulties on the way of elaborating and bringing into force new agreements on PPW, in the second half of 1980s and in the 1990s, the international community focused its efforts on developing confidence-building measures in outer space, mainly within the framework of the Conference on Disarmament (CD) in Geneva, where, in 1985, a Special Committee with a research mandate on PAROS was established. The Special Committee conducted extensive research on possible confidence-building, transparency, predictability measures and measures of control of outer space activities. But there was no way to get them adopted and implemented, because the Special Committee had no negotiation mandate (the US blocked consensus decision on conferring on it a negotiation mandate). In 1994, the Committee suspended its activities (the priority was given to other CD agenda items).

Assessing the international community decades-long efforts to shape the ISL, Russia insists upon further strengthening the OST and the legal regime it created. Any key agreement in such sensitive field as the research and the use of outer space is to be legally binding, because outer space is directly related to strategic stability and international security. No ambiguity in interstate relations concerning outer space is acceptable.

The international campaign to prevent the placement of weapons in outer space is gaining momentum. The existing loopholes in the ISL, as well as the orientation of relevant military and strategic concepts and military and technological capabilities (the revised US National Space Doctrine is a telling example) vividly confirm that the world is facing a totally new and quite dangerous threat of space weaponisation.

It would be reckless to use outer space for placing weapons for the sake of meeting one's own geopolitical goals. The reasons provided in this regard raise significant doubts. In particular, it is being stated that the course towards extensive weaponisation of outer space is unavoidable. It is substantiated as a reaction to new threats, allegedly, generated by WMD and missile technology proliferation; acquisition by some states of the means to disrupt regular operation of space systems, namely by the technologies of "blinding" spacecraft sensors by laser radiation.

Any attempt to put outer space on an equal military footing with land, sea and atmosphere could be extremely dangerous. It is unacceptable to regard space as a "regular environment", where any military operation can be conducted. The mere attempt to counter other states' use of outer space, per se, is a direct violation of the UN Charter Article 2.4. It is noteworthy that the leading role in settling such issues is overtly attributed to ASAT systems of all types: space-, land-, sea- or atmosphere based. The use of strategic offensive weapons is allegedly acceptable to suppress land-based components of an adversary's space systems (flight-control centres, space launchers, stations for command transmission and information reception from spacecraft, etc.). Undoubtedly, such a reckless and irresponsible venture would inevitably result in a military confrontation in outer space with unpredictable consequences for human civilisation.

Russia, together with other Collective Security Treaty Organisation (CSTO), BRICS [Brazil-Russia-India-China-South Africa] and Shanghai Cooperation Organisation (SCO) member-states, as well as all Non-Aligned

Movement (NAM) countries, has been consistently and firmly opposing any attempts of placing weapons in outer space. Presently, one may state with full confidence that the international community is fully aware of the threat of placement of weapons in outer space, and the prevention of such weapons being placed in space becomes one of the most pressing tasks of today.

The current proposals aimed at preventing placement of weapons in outer space contain a wide range of possible measures – starting from transparency and confidence-building measures (TCBMs) up to the preparation of comprehensive international agreements on the prevention of space-based weapons.

In our view, the most acceptable way of reaching the ultimate goal, which is a comprehensive agreement on the non-placement of weapons of any kind in outer space, is a step-by-step approach. Extensive and laborious work with a view of combining efforts of all responsible states is needed. National positions may differ in some aspects, but all responsible states share the same idea – while it is not too late we need to introduce a comprehensive legally binding prohibition on placement of weapons in outer space. At first glance, it may seem that putting such approach into practice is not so hard, for space is still free of weapons. Seemingly, not much is needed – all states should agree, once and forever, to assign a weapon-free status to outer space. Anyway, it is absolutely evident that we have no right to repeat the fatal errors of the past. The US nuclear bombing of Hiroshima and Nagasaki in 1945 heralded a new era of nuclear and missile deterrence, and for a second century now the entire world has been trying hard (but with no real positive prospects in the near future) to effectively address the issue of general and complete nuclear disarmament.

An important prerequisite for gradual progress towards achieving the PAROS/PPW goal lies in ensuring that States possessing military space technologies take no practical steps to put weapons into outer space while the work on new international arrangements is underway. A global moratorium on placement of military assets into outer space could be an effective measure in this regard. Back in 2001, at the 56th UNGA session, Russia declared that it would be ready to take such a commitment immediately if other leading spacefaring nations followed suit.

The UNGA resolution "Prevention of an arms race in outer space" (since 1981 it has been adopted annually almost by consensus) shows optimal ways of implementing a step-by-step approach to the issue of PAROS/PPW. The significance of the resolution is in the international political and legal potential that its key provisions contain. Namely, it reaffirms the important and pressing nature of PAROS; recognises that the prevention of an arms race in outer space would avert a grave danger for international peace and security; underscores the necessity of examination of further measures in the search for effective and verifiable bilateral and multilateral agreements for the purposes of PAROS, including prevention of weaponisation of outer space; reaffirms the importance of CBMs as a means conducive to ensuring the attainment of the objective of PAROS; recognises that the concrete proposals on CBMs could form an integral part of the appropriate international agreements; notes that the legal regime applicable to outer space, by itself, does not guarantee the prevention of an arms race in outer space and that there is a need to consolidate and reinforce it, as well as enhance its effectiveness. Russia fully supports the provisions of the UNGA resolution on PAROS and regularly co-sponsors it.

Strictly following the ideas of PAROS, back in 2004, at the 59th UNGA session, Russia unilaterally came up with an initiative/political commitment not to be the first to place weapons in outer space (no first placement or NFP) and called on all responsible states to follow suit. In 2005, all the CSTO states joined the NFP initiative.

Later on, Russia together with China launched work on an international treaty on the prevention of the placement of weapons in outer space, the threat or use of force against outer space objects (PPWT). In 2008, a Russia-China draft PPWT was tabled at the CD in Geneva (its updated version reflecting comments and proposals by a large number of states was submitted in 2014).

Russia and China also jointly prepared and submitted in 2005 to the UNGA First Committee a draft resolution "Transparency and confidence-building measures in outer space activities" (resolution on TCBMs), which later enjoyed consensus. In accordance with the UNGA resolution 65/68, in 2012-13, a group of governmental experts under Russian Presidency compiled and further developed the existing proposals by States on TCBMs and prepared recommendations on their practical implementation. The TCBMs in outer space are an important element of the verification mechanism of a future PPWT.

Overall, the draft PPWT enjoys an ever increasing international support and remains a key element of PAROS discussions in the CD. Russia, together with China, will continue paying the most careful attention to constructive proposals from interested states regarding further promotion of the PPWT.

However, while the CD is failing to launch talks on the PPWT, Russia is focusing on the implementation of other practical measures designed to address the PAROS/ PPW issue. Namely, priority is given to the globalisation of the initiative/political commitment on NFP. The number of NFP full-fledged participants is increasing each year.

The NFP initiative/political commitment, as an unprecedented TCBM, is becoming an important factor contributing to keeping outer space free of weapons, and, consequently, to the consolidation of international peace, equal and undiminished security for all. The logic here is pretty simple and clear: if all states, for a start, take an NFP political pledge, it will pave the way for a legally binding international treaty banning placement of weapons in outer space and the threat of force against outer space objects. Obviously, in itself, even the forming of an important group of states, which assumed the NFP commitment, is a serious political roadblock on the way to weaponisation of outer space and its turning into an area of armed confrontation.

To sum it up, one needs to admit that, presently, the once prevailing unipolar and bipolar systems of international relations are actively being replaced by an objectively more harmonious polycentric world order, based on the principles of equal and undiminished security for all states without exception, and designed to reflect, not in words, but in deeds, a cultural and civilisational diversity of people living on our planet. Under such circumstances, addressing PAROS/PPWT issues is increasingly getting more importance, and contributes to strengthening global and regional security on a solid foundation of international law.

It is clear that answers to numerous emerging problems could be found only collectively as a result of joint efforts of all responsible states. Any attempt to strengthen one's own security at the expense of security of others holds no water. No national security can be bought at the price of a new mistrust or a brand new insecurity of other nations.

Remarkably, the 1967 OST, which was adopted at the dawn of the space age and for 50 years has been playing its stable role of an impartial regulator in dealing with security problems in outer space activities, in many ways, has predated international development and remains ever more an indispensable pillar of the ISL. Further efforts on consolidation and development of the ISL should be built on the basic principles contained in the OST, such as a free and equal access for all states to the peaceful exploration and use of outer space. We also need to take into account today's realities and rely on the solid background and opinion of such long time major spacefaring nations such as Brazil, India, China and South Africa, as well as on the interests of those countries which are only making first steps in space exploration. There should be no illusion – in a modern world there is no place for utopian and dangerous theories of "exceptionalism and all-permissiveness for the privileged". Only equal cooperation in addressing issues of secure, long-term predictable and sustainable space activities may succeed. The 1967 OST did serve and will continue to serve as a solid base for it.

9

"All's Well that Ends Well": Overview of China's 60-year Space Activity and Space Law

LI Juqian

Space Activity and Standpoint

China's space activity began 60 years ago in 1956 when the country established the Fifth Research Institute under the Ministry of National Defence.[1] This was quite remarkable as well as unique, because the new government had been established just seven years ago in 1949 after three years of civil war and the economy was not strong enough, with the gross domestic product (GDP) amounting to only ¥ 102.8 billion and foreign exchange reserves US$ 0.123 billion.[2] In 1957, the world witnessed a major milestone in the form of the Soviet Union's launch of "Sputnik". This event, preceded by the Korea Peninsula Military Armistice Agreement which was concluded only in 1953, served as a catalyst for the Chinese Space program. The primary reason for developing Space Technology was the need to safeguard its sovereign borders and guard the independence of its country in the era of the Cold War.

China developed space technology and space activities independently and launched the first Chinese manmade satellite "Dong Fang Hong I" into outer space in 1970.[3] China's independent policy for developing space activities was due to the weak industrial and technological basis and special actual situation in the 1950s – it had no choice. After 60 years of development, it can be observed that the independent policy is a success, and progress has been made in all areas including manmade satellites, launching vehicles, manned spacecraft and space station.

Outer Space Treaty and Its Importance

China became a contracting state of the most important United Nations (UN) space treaty, "Treaty on Principles Governing the Activities of States in the Exploration and Use of Outer Space, including the Moon and Other Celestial Bodies" (in short, the Outer Space Treaty or OST), in 1983 – almost 15 years after the Treaty was concluded by the international community. Since then China has fulfilled all the specific obligations in the OST, not just the general legal principles in space law area in the resolutions adopted by the UN General Assembly.

China is a founding member of the UN. As a permanent member of the Security Council, a supporter and beneficiary of the current international legal order, China respects and abides by international law. The Chinese Government has repeatedly declared its adherence to the commonly recognised rules of international law. Needless to say, the rules of international law include the space law treaties where China is a contracting state. Particularly the OST, which has 104 contracting states and is widely recognised as the basic legal document for space activities.[4]

According to the Constitution of China, it is unclear where in the hierarchy of the Chinese domestic legal system the international treaties that China is party to fall under, but it is clear that these treaties are an important part of China's legal system. The power relating to the treaty-making process lies with the three national organs, the State Council, Standing Committee of the National People's Congress (NPC) and President of China. The treaty concluding power exists with the State Council, i.e. the central government; treaty effectiveness is decided by the Standing Committee of the NPC; and the announcement of the treaty effectiveness is made by the President of China.[5] The NPC is the highest authority and national law making organ in China. As the executive organ of highest authority, the Standing Committee of the NPC is also the national law making authority and national law interpretation authority in China.[6] According to Article 67 of the Constitution of China, the Standing Committee of the NPC exercises the power to ratify or abrogate treaties and other important international agreements. While major treaties are dealt with at the NPC level, there are many low level agreements that are concluded by the State Council, and not subject to the approval from the Standing Committee of the NPC.

China joined four of the five UN space treaties – the OST in 1983; Agreement on the Rescue of Astronauts, the Return of Astronauts and the Return of Objects Launched into Outer Space; Convention on International Liability for Damage Caused by Space Objects; and Convention on Registration of Objects Launched into Outer Space, all three in 1988. All four treaties were submitted to the Standing Committee of the NPC for approval.

Further, in case of a conflict between international treaties and domestic law, no law is specifically mentioned regarding the enforcement of these treaties. The main reason behind this is that there is no legislation at the national level in China specifically relating to space activities. However, China is in the process for formulating its national space law, and fulfilling obligations under the space law treaties is one of its main concerns. Besides, the general practice, especially concerning the domestic civil and commercial law, is that an international treaty shall prevail in case of conflicts between the treaty and the relevant domestic law. Several important laws in China have this provision, for example, the Civil Aviation Law, General Principles of Civil Law, Civil Procedure Law, Maritime Environmental Protection Law and Water Law.

An important signal of China's attitude to space law is the fact that China signed the OST in 1983 immediately after adopting its current constitution in 1982. The 1982 Constitution is the basis of the current legal system of China, and is different in many aspects from the constitution and policy during the Cultural Revolution that ended in 1976. Since 1985, China has been cooperating with more than 10 countries, including the US, UK and Italy in the space technology and space activity area.[7] The OST provides a solid legal basis for international cooperation. And owing to a lack of national space law, the OST is very important to China, for cooperation with other countries must be based on a commonly recognised legal system, not a space policy which can be easily changed.

The national space law, currently under preparation, shall include all the important rules in the OST, owing to the importance of the treaty itself and its status in the Chinese legal system. As a matter of fact, international treaty provisions have already been incorporated into two ministerial regulations and regulated in detail so as to follow the requirements in the Registration

Convention and OST. Both the ministerial regulations were enacted in 2001. The *Measures for the Administration of Registration of Objects Launched into Outer Space* relates to the Registration Convention, and the *Interim Measures on the Administration of Licensing the Project of Launching Civil Space Objects* directly relates to the OST requirements, especially Article 6, "The activities of non-governmental entities in outer space, including the Moon and other celestial bodies, shall require authorization and continuing supervision by the appropriate State Party to the Treaty."

PPWT and the Code of Conduct

Only one article, Article 4, in the OST relates to the restriction of weapons in space. It prohibits the placement of weapons of mass destruction on the orbit around the earth and installation of such weapons on celestial bodies. Furthermore, the Moon and other celestial bodies shall be used exclusively for peaceful purposes. The OST is however silent on the issue of conventional weapons placement or tests on the orbit around the earth, e.g. the X-37B, developed by the US and first tested in the space in April 2010.

Since there is no prohibition of such kinds of weapon tests, any country's step forward will inevitably stimulate other countries to do similar tests, which, needless to say, will accelerate the arms race in space.

A treaty of disarmament in outer space is desirable for the stability and peace of the international community. One such meaningful effort has been the Treaty on the Prevention of the Placement of Weapons in Outer Space, the Threat or Use of Force against Outer Space Objects (PPWT), based on a 2002 document related to the placement of weapons in outer space. China and Russia jointly submitted a PPWT draft to the Conference of Disarmament (CD) in 2008. After several debates in the CD, the PPWT draft was updated in 2014. Unfortunately, the treaty negotiations are yet to be concluded. Nonetheless, China and Russia have the common understanding for advancing the PPWT as a treaty of legally binding force.[8] The two countries also signed the joint statement on strengthening global strategic stability on June 25, 2016 as part of the common effort towards making the PPWT legally binding.[9] Taking into consideration the US President-elect Donald Trump's antagonising attitude to China regarding the Taiwan issue, it looks much more difficult for the PPWT to conclude soon.

Apart from the draft PPWT, does the international community have other ways to reduce the risk of an arms race in outer space? The European Union's

Code of Conduct (CoC) for Outer Space Activities is one of the ways. As a kind of a soft law, before the rules can be accepted as treaty, it is meaningful to have the CoC as a regional code in Europe or other documents passed by the UN General Assembly or other organs as a general code. The formation of soft law is usually easier than the treaty-making process, and also can be used to forge consensus to facilitate treaties in the future. Nevertheless, the CoC itself is not a legally binding force, reducing its effectiveness.

Concluding Observations and Concerns

China has been developing space activities and technology independently, but is also willing to cooperate with any country, especially the neighbouring countries via the Asia Pacific Space Cooperation Organisation (APSCO), which headquarters in Beijing. Moreover, China is committed to the peaceful use of outer space.

Besides the prevention of weaponisation of space, unlawful appropriation is also an area of concern for China. In 2005, the Chinese Government fined a Beijing businessman for selling land on the Moon, and the court upheld the government's sanction. However, if a foreigner or foreign country law has the possibility to appropriate resources in outer space, for example, the US Space Resources Exploration and Utilization Act of 2015, would this be a violation of international law, or is that permissible within the meaning of "free exploration and use" by all states in the OST?

Further, mitigation of space debris is a key area of concern for China. The collision of US and Russian satellites in 2009 warned the world that space debris was far more dangerous than we had imagined. Probably the active way for mitigation of space debris is to remove it from the orbit around the earth. The successful launch of a new generation space rocket in 2016, a first from the Wenchang launch site in China, is a real step towards removing space debris by robotic arms. This shows that long-term sustainable use of outer space is not only an agenda on paper, but is in reality possible too.

ENDNOTES

1. See Xinhua News Agency, http://news.xinhuanet.com/mil/2005-07/31/content_3279519.htm (Accessed December 20, 2016). See also LI Juqian, "Progressing towards New National Space Law", *Journal of Space Law*, 35, p. 440.
2. 1956 年经济数据
3. White Paper 2000, *China's Space Activities.*

4. See COPUOS document, A/AC.105/C.2/2016/CRP.3, "Status of International Agreements Relating to Activities in Outer Space as at 1 January 2016 ", p. 10.
5. See the Constitution of the People's Republic of China 1982, Article 67, Article 81, Article 89.
6. Legislation Law of the People's Republic of China 2000, Article 7.
7. White Paper of China's Space Activity, 2000.
8. Ambassador FU Cong's statement, at http://www.china-un.ch/chn/hyyfy/t1388472.htm (Accessed December 20, 2016).
9. For the whole document, see http://news.xinhuanet.com/politics/2016-06/26/c_1119111895.htm (Accessed December 20, 2016).

10

India and the Outer Space Treaty

Kumar Abhijeet

Introduction

Since the inception of its space programme, India has been committed towards its international obligations flowing from the 1967 Outer Space Treaty (OST)[1] and other international agreements. India is a party to the OST, and with more than five decades of successful, vibrant and robust state-sponsored space programme, not a single occasion has been witnessed where any of its space activity has been in conflict with the OST. Such an unblemished record is a result of India's commitment to international obligations and the realisation that space is to be used for societal benefits. In the words of Dr Vikram Sarabhai – father of the Indian space programme,

> There are some who question the relevance of space activities in a developing nation. To us, there is no ambiguity of purpose. We do not have the fantasy of competing with the economically advanced nations in the exploration of the Moon or the planets or manned space flight. But we are convinced that if we are to play a meaningful role nationally, and in the community of nations, we must be second to none in the application of advanced technologies to the real problems of man and society.

So far space activities in India have been the prerogative of the government alone and compliance with international obligations has not been that problematic even in the absence of express legislation. However, as India gets ready to open up the space sector for privatisation, things will no more remain same, particularly as the OST imposes liability for all non-governmental activities too on the respective state. Nonetheless, today, private actors have the technological and financial capacity to undertake hazardous risky activities in outer space, and now India is positively looking to involve private entities for space activities. Against the backdrop of private sector involvement, this chapter examines the nature of obligation as levied by the OST, weighs the existing domestic measures to implement the Treaty obligation and proposes the future course of action.

Obligations Levied by the OST

Freedom in Outer Space

Article 1 of the OST grants three kinds of freedom in outer space: (i) freedom of exploration and use of outer space including the moon and other celestial bodies;[2] (ii) freedom of access to all areas of celestial bodies;[3] and (iii) freedom of scientific investigation in outer space including the moon and other celestial bodies.[4]

Hobe has commented that freedom here connotes that any entity is free to explore or find out possible use of outer space without any permission from any other state.[5] A state is free to take any space activities including economic activities, and even profit from these activities.[6] This freedom is not restricted only to the government but also available to non-governmental entities and individuals via Article VI[7] of the OST.

The freedom in outer space including the moon and other celestial bodies is not an absolute freedom. There are inherent limitations on freedom within Article 1, and some are also expressed in the other Articles of the OST. Paragraph 1 (Article 1) says that freedom shall be exercised for the 'benefit and interests of all countries' and that outer space shall be the 'province of mankind'. It reminds the states that outer space is not under the jurisdiction of specific states and therefore an activity carried out in outer space and on celestial bodies may not be undertaken for the sole advantage of states.[8] The freedom is to be exercised in a non-discriminatory manner, on the basis of equality

The limitations stated outside Article 1 include the non-appropriation principle,[9] applicability of international law and Charter of the United Nations,[10] limited military use,[11] international responsibility for national activities[12] and avoidance of harmful contamination[13].

Non-appropriation Principle

Article 1 guarantees freedom of exploration and use of outer space including the moon and other celestial bodies, but Article 2 ensures that outer space "is not subject to national appropriation by claim of sovereignty, by means of use or occupation, or by any other means."[14] The 'non-appropriation' principle is the fundamental rule regulating the exploration and use of outer space that aims to protect outer space from the possibility of conflict driven by territorial or colonisation-driven ambition.[15] It prohibits sovereign or territorial claims to outer space. No amount of the use or occupation of outer space will ever suffice to justify a claim of ownership rights over the whole or any part of outer space, including the moon and other celestial bodies.[16] Despite few instances[17] of unsuccessful attempts to challenge the non-appropriation principle, it can be concluded that the principle is not only a norm of international space law but also has attained the customary law status.[18]

Space Exploration in Accordance with International Law

Article III of the OST widens the ambit of legal prescription to space activities. It makes international law, including the Charter of the United Nations, applicable to space activities. Judge Lachs has expressed that "the obligation to confirm with the Charter of the United Nations implies not only the application of provisions of international law as defined by it but also all those that have grown as a result of the further development of the United Nations and subjected to a new and more up-to-date interpretation". Thus, even though there has been no new treaty governing space activities besides the 1979 Moon Agreement, the domain of international space law is continuously evolving and new principles, treaties become applicable to space activities as well.

Peaceful Use of Outer Space

Article IV of the OST has strived to limit use of space for peaceful purposes only. It prohibits placing of nuclear weapons or weapons of mass destruction (WMD) in orbit around the Earth.[19] The establishment of military bases, installations and fortifications, the testing of any type of weapons and the conduct of military manoeuvres on celestial bodies is forbidden. The moon and other celestial bodies are to be used exclusively for peaceful purposes.

At times dual use of space technology makes it difficult to ascertain the exact nature of activity. For example, a space object using nuclear fuel could also be used as a weapon, making it difficult to define what constitutes a 'nuclear weapon'. Moreover, Article IV does not prohibit placing of conventional weapons or military satellites in orbit.[20] The paradox of Article IV is that though it expressly prohibits placing of nuclear weapons or WMD in orbit or on celestial bodies but it does not prohibit actual use of such weapon. Therefore, outer space may be used as a transit area for weapons aimed at and used on Earth, and in theory WMD may be launched and initiated in space.[21] Similarly, it does not prohibit Anti-satellite weapons. Above all, peaceful use has not been defined which gives scope to non-aggressive military use. Despite the limitations of Article IV, it has been able to considerably restrict use of space to peaceful purposes. Disarmament in space is a major agenda for the United Nations and international cooperation can only enable preservation of space.

Astronauts as Envoys of Mankind

Article V of the OST carries a humanitarian element and concerns the effective protection of people involved in the exploration of outer space on behalf of all mankind.[22] All astronauts are to be treated as "envoys of mankind in outer space" and in the event of accident, distress, or emergency landing states are under an obligation to render all possible assistance to the astronauts of other states and they must be safely and promptly returned to the state of registry of their space vehicle.[23] The 1968 Rescue Agreement[24] further elaborates this responsibility.

The benefits of Article V are for astronauts who are authorised trained state personnel. But in the light of space tourism, it is essential to start a discourse as to whether the same benefits can be granted to space tourists or something else needs to be done.[25]

Responsibility and Liability of Space Objects

It is the responsibility of states to assure that their national activities are carried out in conformity with the OST.[26] In case of damage caused by space objects, the launching state bears international liability to compensate for the damage caused.[27] (These two aspects have been elaborated later in this chapter under the section, Non-governmental Entities and the OST)

Avoidance of Harmful Contamination

Article IX of the OST may be considered as the basis for environmental protection of outer space and its preservation for peaceful use. Activities in outer space are ultra-hazardous possessing the risk of harm to both terrestrial as well as outer space environment. States must "conduct exploration of outer space, including the Moon and other celestial bodies, in such a way so as to avoid their harmful contamination and also adverse changes in the environment of the Earth resulting from the introduction of extraterrestrial matter and, where necessary, [to] adopt appropriate measures for this purpose". Moreover, where a state has reason to believe that an activity or experiment planned by it or its nationals in outer space, including the Moon and other celestial bodies, would cause potentially harmful interference with activities of other states parties in the peaceful exploration and use of outer space, including the Moon and other celestial bodies, it must undertake appropriate international consultations before proceeding with any such activity or experiment.

Recent developments have suggested that the OST has its own limitations and states must strive to overcome them. But, despite its limitations, the OST gives sufficient legal framework to restrict nations to peaceful use of outer space.

Space Law in India

From the preceding section it is clear that the OST imposes enormous responsibilities upon state parties. India being signatory to and having ratified the OST, the current section weighs the existing domestic measures to implement the Treaty obligations.

It is constitutionally mandated that India shall endeavour to 'promote international peace and security' and 'foster respect for international law and treaty obligations'.[28] Obligations flowing from international treaties can be implemented either by way of legislative action[29] or through executive action[30] vested in the President of India. As of today no specific space legislation has been enacted by the Government of India to implement the obligations of the OST, conversely the obligations are implemented by executive action.[31] Article 77 of the Constitution of India lays the basic principle for conduct of business of the government. All executive action of the government is to be taken in the name of the President of India. For the more convenient transaction of the business of the government, and for the allocation among Ministers of the said business the President can make rules.[32] Pursuant to this rule-making power, the Government of India (Allocation of Business) Rules, 1961 has been enacted. The First Schedule, Rule 2(45) of the Allocation of Business Rules accommodates the Department of Space (DOS), and the Second Schedule allocates powers and function to the DOS in all matters relating to space science, space technology and space applications including international relations in matters connected with space.

In addition to the Allocation of Business Rules, the 1997 Satellite Communications (SATCOM) Policy and the 2011 Remote Sensing Data (RSD) Policy compliments the legal regime for space activities in India. The fundamental aim of the SATCOM Policy is to develop a healthy and thriving communication satellite and ground equipment industry as well as satellite communications service industry; use and further development of the capabilities built in India in the area of satellites, launch vehicles and ground equipment design and sustaining these capabilities; encouraging the private sector investment in the space industry in India and attracting foreign investments in this area.[33] The 2011 RSD Policy contains modalities for managing and permitting the acquisition and dissemination of remote sensing data in support of developmental activities.[34]

Even though space law in India is not yet well developed, but to date the DOS has been diligent enough in ensuring that space activities have been in accordance with the OST. The DOS Citizen's Charter offers a reference point to the understanding of India's vision and objectives for the exploration and use of outer space. It has the primary objective of promoting development and application of space science and technology to assist in all-round development of the nation.[35]

Non-governmental Entities and the OST

Although at the time of drafting of the OST, all space activities were completely in governmental domain but the drafters anticipated that there could be a time when private entities would participate. The OST is the first international agreement that specifically addresses non-state actors.[36] As per Article VI of the OST, activities of non-governmental entities in outer space shall require authorisation and continuing supervision by the appropriate state.[37] Furthermore, states bear international responsibility not only for governmental agencies but also for their non-governmental entities.[38] It follows that, whereas the non-governmental entity reaps the freedom in outer space, the respective state carries the responsibility to ensure that activities of non-governmental entities are in accordance with the OST. Failure to effectively discharge this international responsibility may even make the states liable to pay for damages caused by space activity.[39] This liability is unlimited in time, amount and quantum.[40]

Article VI of the OST only demands 'authorisation and continuing supervision' which can be achieved by any means not necessarily through legislation, but, as discussed earlier, the obligations flowing from the OST are so enormous and complicated that it is in the interest of respective states to enact legislation.[41] Above all, 'public liability for private activities'[42] is a strong incentive for states to authorise their space activity through legislation.[43] Thus, the principle of 'authorisation and continuing supervision' is considered the starting point for a discussion on national space legislation.[44] Pursuant to Article VI many spacefaring nations have enacted a domestic legislation enabling private sector participation as well as shielding governmental liability.

Article VIII of the OST confers jurisdiction and control over objects launched into space and over any personnel on the state on whose registry such space objects are carried: "Ownership of objects launched into outer space, including objects landed or constructed on a celestial body, and of their component parts, is not affected by their presence in outer space or on a celestial body or by their return to the Earth. Such objects or component parts found beyond the limits of the State Party to the Treaty on whose registry they are carried shall be returned to that State Party, which shall, upon request, furnish identifying data prior to their return."[45]

Thus, registration of space objects benefits states in three ways: (i) In case of violation of the original authorisation conditions, the concerned state need not do anything extra to take over jurisdiction and control of such space objects as it is the states that bear obligation to register their space objects and not their private entities. (ii) It prohibits any unauthorised transfer of operation/in orbit sale of space objects, safeguarding the financial interest of states.[46] (iii) It shall serve as an evidence for claim of return of space objects and personnel found beyond the national territory.

With India poised to open up the space sector to private players, it is important to reiterate that it bears international responsibility for the activities of its private entities. In other words, it is responsible for ensuring that all private space activities are in accordance with the OST. As a launching state, it bears international liability for damages caused by space objects and also the obligation to register its space objects. Moreover, under no circumstances can the launching state escape this liability because 'once a launching State forever a launching State'[47]. Hence, the launching state needs to prescribe a mechanism for recourse, should it incur any liability for private entities. Thus, in the given context, India's international treaty obligations especially Articles VI, VII and VIII of the OST make national space legislation of paramount concern.[48]

Conclusion

The Indian space programme has successfully survived for five decades without domestic legislation and completely in compliance with the OST. But with private sector involvement, compliance with the OST may not be that easy in the absence of legislation. As discussed in this chapter, there exists a wide gap between the requirements of the OST and the existing legal regime for space activities in India. Realising the wide gap and the enormous international responsibility, the Government of India has enacted a draft legislation that is being scrutinised by an inter-ministerial body.[49] Thereafter the draft will be in public domain. Hopefully, the legislation would have addressed issues of authorisation, continuing supervision, registration of space objects, liability for damages, environmental safeguards and other legal issues flowing from the OST.

ENDNOTES

1. See United Nations Office for Outer Space Affairs (UNOOSA), "Treaty on Principles Governing the Activities of States in the Exploration and Use of Outer Space, Including the Moon and Other Celestial Bodies", opened for signature in Moscow, London and Washington in January 1967, and entered into force in October 1967, at http://www.unoosa.org/oosa/en/ourwork/spacelaw/treaties/introouterspacetreaty.html.
2. Ibid., Article 1, Paragraph 1.
3. Ibid., Article 1, Paragraph 2.
4. Ibid., Article 1, Paragraph 3.
5. Stephan Hobe, "Article 1", in Stephan Hobe et al. (eds.), *Cologne Commentary on Space Law*, Volume 1, *Outer Space Treaty*, Carl Heymanns Verlag, Cologne, 2009, p. 34.
6. Ibid., p. 35.
7. UNOOSA, No. 1, Article VI: "States Parties to the Treaty shall bear international responsibility for national activities in outer space, including the Moon and other celestial bodies, whether such activities are carried on by governmental agencies or by non-governmental entities, and for assuring that national activities are carried out in conformity with the provisions set forth in the present Treaty. The activities of non-governmental entities in outer space, including the Moon and other celestial bodies, shall require authorization and continuing supervision by the appropriate State Party to the Treaty."
8. Stephan Hobe, No. 5, p. 37.
9. UNOOSA, No. 1, Article II.
10. Ibid., Article III.
11. Ibid., Article IV.
12. Ibid., Article VI.
13. Ibid., Article IX.
14. Ibid., Article II.
15. Steven Freeland and Ram Jakhu, "Article 2", in Stephan Hobe et al. (eds.), No. 5, p. 48.
16. Ibid., p. 53.
17. In 1976, a number of equatorial States – including Brazil, Columbia, the Congo, Ecuador, Indonesia, Kenya, Uganda and Zaire issued the Bogota Declaration claiming state sovereignty

over parts of Geostationary Orbit; See Ram Jakhu, "The Legal Status of the Geostationary Orbit", *Annals of Air and Space Law* 1982, p. 333; Thomas Gangale, "Who Owns the Geostationary Orbit?", *Annals of Air and Space Law*, 2006, p. 425.

In 2001, the Gregory W. Nemitz claim to Aestorid 433 was rejected. See Nemitz v. US, Slip Copy, 2004 WL 316704, D. Nev., 2004 (April 26, 2004) and Nemitz v. NASA, 126 Fed. Appx. 343, 9th Circ. (Nev.), February 10, 2005. The legal validity of sale of plots on the Moon by private companies was considered illegal. See Statement by the Board of Directors of the International Institute of Space Law on Claims to Property Rights Regarding The Moon and Other Celestial Bodies (2004). Also see Steven Freeland and Ram Jakhu, No. 15, p. 55.

18. Ibid. (Steven Freeland and Ram Jakhu).
19. UNOOSA, No. 1, Article IV.
20. Kai-Uwe Schrogl, "Article IV", in Stephan Hobe et al. (eds.), No. 5, p. 78.
21. Ibid.
22. Ben Cheng, "The Legal Status of Astronauts", in *Studies in International Space Law*, Clarendon Press, Oxford, 1997, p. 457.
23. UNOOSA, No. 1, Article V.
24. India has signed and ratified the 1968 Rescue Agreement.
25. F.G. von der Dunk and Gerardine Meishan Goh, in Stephan Hobe et al. (eds.), No. 5, pp. 94-102.
26. UNOOSA, No. 1, Article VI.
27. Ibid., Article VII.
28. Article 51 of the Constitution of India.
29. Article 253 of the Constitution of India – "Notwithstanding anything in the foregoing provisions of this Chapter, Parliament has power to make any law for the whole or any part of the territory of India for implementing any treaty, agreement or convention with any other country or countries or any decision made at any international conference, association or other body."
30. Article 53(1) of the Constitution of India – "The executive power of the Union shall be vested in the President and shall be exercised by him either directly or through officers subordinate to him in accordance with this Constitution."
31. Ranjana Kaul, "National Space Legislation: A Blueprint for India", *Proceedings of the ISRO-IISL Space Law Conference 2005, Bringing Space Benefits to the Asia Region 26-29 June 2005*, Allied Publishers, Bangalore, pp. 2-14.
32. Article 77(3) of the Constitution of India.
33. http://www.isro.gov.in/indias-space-policy (Accessed October 2016).
34. http://dos.gov.in/sites/default/files/RSDP-2011.pdf (Accessed October 2016).
35. http://dos.gov.in/node/25 (Accessed October 2016).
36. Stephan Hobe, No. 5.
37. UNOOSA, No. 1, Article VI, Sentence 2.
38. Ibid., Sentence 1.
39. Ibid., Article VII: "Each State Party to the Treaty that launches or procures the launching of an object into outer space, including the Moon and other celestial bodies, and each State Party from whose territory or facility an object is launched, is internationally liable for damage to another State Party to the Treaty or to its natural or juridical persons by such object or its component parts on the Earth, in air space or in outer space, including the Moon and other celestial bodies."

    Article VII has further been elaborated by the 1972 Liability Convention to which India has both signed and ratified.

40. Armel Kerrest, "Sharing the Risk of Space Activities: Three Questions, Three Solutions", in Karl-Heinz Böckstiegel (ed.), *Project 2001 – Legal Framework for the Commercial Use of Outer Space*, Carl Heymanns Verlag, Cologne, 2002, p. 136.
41. Michael Gerhard, "Article VI", in Stephan Hobe et al. (eds.), No. 5, p. 119.
42. Frans G. von der Dunk, "The Origins of Authorisation: Article VI of the Outer Space Treaty and International Space Law", in F.G. von der Dunk (ed.), *National Space Legislation in Europe: Issues of Authorisation of Private Space Activities in the Light of Developments in European Space Cooperation*, Martinus Nijhoff Publishers, Leiden, 2011, pp. 3-28.
43. Irmgard Marboe, "National Space Law", in Frans von der Dunk and Fabio Tronchetti (eds.), *Handbook of Space Law*, Edward Eligar Publishing, Cheltenham, 2015, pp. 127-204.
44. Michael Gerhard, No. 41, p. 117.
45. Complementary to Article VIII OST, the 1975 Registration Convention was enacted to impose obligation upon States to register their space objects launched into outer space in the United Nations registry as well as national registry. India has signed and ratified the Registration Convention.
46. Michel Gerhard, "Transfer of Operation and Control with Respect to Space Objects – Problems of Responsibility and Liability of States", *German Journal of Air and Space Law*, 4, 2002, pp. 571-581.
47. Henry Hertzfeld and Frans G. von der Dunk, "Bringing Space Law into the Commercial World: Property Rights Without Sovereignty", *Chicago Journal of International Law*, 5, 2005, pp. 89-90; also see Frans G. von der Dunk, "The International Law of Outer Space and Consequences at the National Level for India: Towards an Indian National Space Law?", *Indian Yearbook of International Law and Policy*, 2009; Frans G. von der Dunk, No. 42.
48. Ibid. (Frans G. von der Dunk, 2009), pp. 135-163.
49. Kalyan Ray, "First Indian Space Law Being Vetted by Military", *Deccan Herald*, February 29, 2016.

11

Japan, Koreas and the Outer Space Treaty

Munish Sharma

The politico-military and security dynamics between the triad of Japan, the Democratic People's Republic of Korea (DPRK) and Republic of Korea (ROK) has played an important part in shaping their respective space programmes. The geographical proximity of this triad and the complex security imperatives arising out of the military and technological developments compel an analysis of their space programmes collectively from the standpoint of the Outer Space Treaty (OST). The regional dynamics among and between the three nations, along with their technological interactions or exchanges with the major powers of the world have had a deep impact on the security imperatives in such a way that the military aspect of space programme development is dominant as compared to the civilian aspect. Moreover, given the dual-use nature of space technology, especially rocket propulsion, the development of space programme in the Korean Peninsula has been military driven. These factors have severe implications for the exploration of space for peaceful purposes, the key pillar of the OST.

The OST has stood the test of time since it came into existence in 1967. It has been a landmark treaty in this domain along with a number of international agreements and conventions on space. However, it should be noted that the DPRK became a signatory to the OST quite late, in 2009, while Japan and the ROK have been party to the treaty since 1967. The domestic practices and policies governing the development of space technologies in these respective countries have since then been measured with the prism of principles laid down by the OST.

The DPRK and the OST

The DPRK signed the OST in March 2009.[1] In April, the same year, the DPRK attempted a satellite launch, and the failed trial drew a lot of international criticism.[2] The DPRK Government asserted that the launch was justified as per a principle of the treaty, which states that "space shall be free for exploration and use by all states without discrimination of any kind".[3] With continued attempts the DPRK managed to place a satellite successfully in orbit in December 2012.[4]

The international assessment (China, the US, Japan and the ROK) with regard to the DPRK's attempts to launch satellites terms them to be "aimed at developing an inter-continental ballistic missile capable of striking the US"[5], however the DPRK Government claims that these launches are meant for scientific studies. South Korea, Japan and the US called the February 2016 satellite launch (Kwangmyongsong-4) to be "unforgivable", "intolerable" and a "major provocation".[6] The DPRK argued that this launch was not a violation of the OST as the treaty permits the states party to treaty "free exploration and use of space by all states without any discrimination".[7] The DPRK Government has termed the Kwangmyongsong-4 as an "earth observation satellite" with the purpose to "monitor the weather and to map the location of natural resources and forests".[8] As per the treaty terms, the DPRK's ambitions and attempts to launch satellites are legitimate, if their satellites find applications in security or developmental objectives.

International security concerns regarding the DPRK primarily arise out of its nuclear weapons programme. The United Nations Security Council (UNSC) has adopted five major resolutions since 2006 that impose and strengthen sanctions on the DPRK for continuing to develop its nuclear weapons programme, and call on the DPRK to dismantle the programme and refrain from ballistic missile tests.[9] The DPRK is prohibited to undertake such satellite launches as per the UNSC Resolutions 1718 (passed in October 2006) and 1874 (passed in June 2009), because the technology for satellite launch vehicle has dual-use application in the development of ballistic missiles.[10] With advancement of the DPRK's nuclear programme, since its withdrawal from the Nuclear Non-proliferation Treaty (NPT), access to missile technology undermines the security of many nations in adversarial relations.

The DPRK has weak credentials in its adherence to international treaty mechanisms. The DPRK became a party to NPT in 1985 as a non-nuclear-weapon state. Citing Article X of the NPT, it withdrew from the treaty in 2003 to avoid the International Atomic Energy Agency (IAEA) demands to inspect its efforts to produce nuclear weapons.[11] The OST (Article XVI) also provisions any state party to the treaty to withdrawal from the treaty.[12] As a signatory to the OST, the DPRK has the right to launch satellites into orbit for earth observation or any other peaceful action. But, it is noteworthy that technology development in the DPRK is primarily driven by military ambitions. Over the last two decades, the DPRK Government has put thrust on its nuclear weapon and missile development programmes as a strong pillar of its national defence and pride.

The domestic economic conditions have rendered the DPRK incompetent vis-à-vis its neighbours in conventional forces. This has led Pyongyang to expand and strengthen its asymmetric capabilities.[13] The investments of the DPRK in the space domain and the related policy matters are influenced by military interests. Therefore, its space programme is possibly an offshoot of its ballistic missile development programme, and it may not have any clear or known civilian applications. International security concerns are also legitimate as the DPRK is prohibited to develop or test ballistic missiles and nuclear weapons under the UNSC resolutions. Although the DPRK will attract further sanctions as it conducts satellite launches, advertising a successful launch is an opportunity for the government to unite its citizens and display it as a matter of national pride.

The advancements of the DPRK in the nuclear and missile arena alter the regional security calculus as well. With technology-driven economics – Japan and the ROK – in the vicinity, internationally isolated and technologically stripped DPRK banks heavily on its strategic capabilities and its ties with China and Pakistan.[14] If the DPRK's claims about the satellite applications are false and the alleged technological support from China or Pakistan is true, all the satellite launch activities of the DPRK seriously undermine the OST, space security and regional/international peace and stability.

The ROK and the OST

The ROK is the fourth-largest economy of Asia and one of the G-20 major economies of the world.[15] As a developed, technology-driven and heavily industrialised nation, the ROK is keen to invest in space technologies to reap the benefits for socioeconomic objectives. The ROK embarked on the journey of space exploration in the late 1980s. In conjunction with its satellite programme, the Korea Aerospace Research Institute (KARI) also began experiments in sounding-rocket technology in the 1990s. Due to missile development restrictions beyond a limit, KARI concentrated on building a network of communications and remote-sensing satellites.[16] However, the advancing ballistic missile programme of the DPRK and its attempt to launch a satellite using Taepodong I in 1998 was a direct military threat to the ROK, which challenged its technological prowess and national security. Facing initial failures in the development of an indigenous satellite launch capability, it has succeeded in satellite design and manufacturing, with 12 satellites in orbit and some of them meant for military communications purposes. The earth observation satellites have civilian uses, but their utility in defence and security operations cannot be completely ruled out.

The ROK signed the OST in January 1967 and submitted the ratification in October, the same year.[17] Space is integral to the development and prosperity of a nation state. As a signatory to the OST, the ROK is bound to restrict its exploration practices to peaceful purposes and to strengthen international peace or its own national security. However, the triad of Japan, the DPRK and ROK shares a fragile geopolitical environment where advances in strategic domains are monitored closely. Any assessment of the ROK's investments in space exploration has to be based on the fragile geopolitical realities of the region, with the DPRK at one end as a credible threat and the US as a net security provider. With the changing security dynamics in the Korean Peninsula, the effectiveness of South Korea's traditional deterrence and defence strategies has eroded.[18] The DPRK and ROK as political and military rivals have their own independent threat perceptions as they remain in a state of war with heavily militarised borders.

The ROK and Japan ensure sustained and active military presence of the US in East Asia, which has raised concerns in security establishments across the DPRK and China. At the end of the Korean War, the US and ROK signed a Mutual Defense Treaty in 1953, which still endures as the foundation of a comprehensive security alliance.[19] In accordance with the treaty, the US has maintained military presence in the ROK in support of its commitment to help the country defend itself against external aggression.[20] The US extends its "nuclear umbrella" to the ROK,[21] and the DPRK remains a common threat. On the contrary, the DPRK alleges the ROK to be "a mouthpiece, a puppet of the United States" having no jurisdiction or control of its own affairs. It dares the ROK to stop towing the hostile policy of the US towards the DPRK and come out of the US military umbrella to normalise the security situation across the Korean Peninsula.[22]

Further, sceptics in China, too, are increasingly questioning the US military ambitions in the region, particularly the "nuclear umbrella" and Ballistic Missile Defence, which according to Chinese analysts nullify China's strategic nuclear deterrent. China has raised concerns about the deployment of Terminal High-Altitude Area Defence (THAAD) system, too, which could allow the US to monitor airspace deep into Chinese territory and upset the "strategic balance".[23]

The ROK maintains the position that "as a state party to all major conventions on outer space, it has carried out all of its outer space activities in a peaceful, transparent and safe manner in accordance with the relevant international norms".[24] It has also signed a Civil Space Framework Agreement with the US in 2016. Besides Russia and the US, KARI cooperates with members of the European Space Agency, Ukraine, India and Japan.[25] The ROK understands that space-based technologies are important for national security missions, such as space-based reconnaissance, particularly when there is a direct threat of missile attack from across the border which could even carry a nuclear warhead. The Mutual Defense Treaty provides the desired security assurance to the ROK from its authoritarian rival. Despite the assurances, the ROK has undertaken an indigenous space development programme, which has vast civilian applications and serves the desired military purposes as well, such as reconnaissance and surveillance for early warning. If these stated applications strengthen its security and ensure stability in the region, the activities are in line with the principles of the OST. More than the domestic policies of the ROK, it is the US whose foreign policy and military ambitions are likely to alter the calculus. However, any disruption in this fragile ecosystem, where different players are at different pedestals of space and missile development, will not just trigger an irreversible competition for strategic advantage but will also impair the international efforts to restrict space exploration to peaceful purposes only.

Japan and the OST

Japan was the first Asian country to place a satellite into orbit in February 1970. Japan was also the third country to succeed in placing a satellite in the geostationary orbit and sending spacecraft to Moon and Mars. Japan is an important participant in the International Space Station (ISS), with its own orbiting laboratory, Kibo. In the last five decades of space exploration, as a leading spacefaring nation, Japan has significant contributions in the space arena with satellites for remote sensing, communication and meteorology applications.

Japan became signatory to the OST in January 1967 and it maintains the position that "placement of weapons of mass destruction in outer space or orbit is absolutely forbidden", as per the OST. Japan's space exploration practices were limited by the 1969 resolution of its lower house of parliament to use space exclusively for "peaceful purposes"[26] only and excluded the Self-Defense Forces (SDF) from directly engaging in any activity related to satellites meant for communications, global positioning or reconnaissance.[27] Restrictive defence policies of Japan in the space domain had prevented Japan from developing advanced reconnaissance satellites which could provide high resolution imagery. The Basic Space Law of Japan, enforced in 2008, lifted the four-decade-old restrictions on the military use of satellites. The law empowers Japan to take necessary measures to promote Space Development and Use to ensure international peace and security as well as to strengthen its national security,[28] within the framework of the pacifist principles of the Constitution of Japan.[29]

The new space law has shaped and guided Japan's space policy in the last one decade. The law states that "the SDF can manufacture, possess and operate its own satellites to support its terrestrial military operations, including Ballistic Missile Defence. Immediate candidates for SDF procurement would be satellites for reconnaissance, early warning and tracking and communications – all to enhance BMD capabilities".[30] Again, the changes in domestic space policy are security driven, deeply rooted in the geopolitics of East Asia.

The most credible and immediate threat to Japan arises from the DPRK, as it advances both its nuclear weapon programme and ballistic missile development. Under these circumstances, Japanese security concerns are legitimate, particularly when Japan has already witnessed a shocking breach of its airspace as the DPRK test fired its intermediate range ballistic missile Daepodong-1 in 1998.[31] This gave impetus to Japan for rapid deployment of reconnaissance satellite network for early warning system against ballistic missile launches.[32] However, the changes in Japan's posturing did not deter the DPRK and it conducted another round of ballistic missile tests in the Sea of Japan in July 2006.[33] The Japanese Ministry of Defense terms the 2006 missile and nuclear tests as "serious and acute threats" to Japan's security and "peace and stability" in East Asia.[34] Moreover, China's military growth and advancements in the field of space exploration, especially in the military domain is more complex and ambiguous threat to Japan, also being a long term challenge. China's 2007 anti-satellite (ASAT) weapon test provoked international condemnation and a critical response from top members of

Japan's leadership.[35] China's ASAT capabilities are a clear threat to Japanese space-based assets, under the pretext of existing and unresolved territorial disputes in the East China Sea.

Post-World War II, the US-Japan alliance has defined the security perceptions of East Asia. The Treaty of Mutual Cooperation and Security between the two nations grants the US the right to military bases on the archipelago in exchange for a US pledge to defend Japan in the event of an attack. The DPRK and China, in loggerheads with the US which provides security umbrella to Japan, have their own suspicions and sceptics over this security alliance, fuelling a military competition which percolates into the space domain. The missiles of the DPRK are capable of reaching Japan and threaten the US military bases there. As a response to these mutual threats, the US and Japan have invested heavily in Ballistic Missile Defence system, which alters the security calculus in East Asia.

The security measure as part of the US-Japan alliance make it imperative for Tokyo to enhance and strengthen its space capabilities. Japan has acquired its impressive capabilities in full view of a pacifist public and under constitutional constraints.[36] In addition, the current trend of space development is also perceived in Japan as a race to prestige and technological superiority.[37] Japanese Ministry of Foreign Affairs also maintains the position that the use of space by the SDF does not violate the principles of the "peaceful use of outer space" and considers its space programme to be significant to its security and position as one of the world's leading countries in space development.[38] The dual-use nature of satellite launch vehicles and nuclear technology always leaves a window of opportunity open for military uses of these strategic technologies. Japan with its technological prowess is by no means inferior to any of its adversaries in the region. Japan's military options for space exploration were limited due to its domestic laws. It is quite evident that domestic laws do not restrict Japan from exercising its right to secure its interests and strengthen its national security in tandem with the changing geopolitical environment. Further race towards ballistic missiles, nuclear or ASAT weapons and missile defence systems will push the militarisation and arms race into space, subverting the principles and tenability of the OST.

Conclusion

The relations between the triad of the DPRK, ROK and Japan in East Asia have unique geopolitical underpinnings, clearly influenced by the presence of regional and extra-regional major powers. The fragile Offence-Defence balance could be disrupted with induction of missile defence which is perceived as a threat to strategic deterrent. Although it is unanimously agreed that the use of space should be strictly restricted to peaceful purposes, which are fundamentally civilian in nature, but there certainly are some limited military applications which not just ensure national security but also strengthen international or regional peace. From the vantage point of the OST, the three nation states under analysis have undertaken their respective space development programmes in accordance with the terms of the treaty. Given the opacity in the DPRK regarding its nuclear weapons programme, the uncertainties owing to the authoritarian regime, the active military presence of the US as part of its Asian pivot and expanding strategic military ambitions of China will put this five-decade-old treaty to test.

ENDNOTES

1. "Treaty on Principles Governing the Activities of States in the Exploration and Use of Outer Space, Including the Moon and Other Celestial Bodies", United Nations Office for Disarmament Affairs, at http://disarmament.un.org/treaties/t/outer_space (Accessed January 10, 2017).
2. Choe Sang-Hun and David E. Sanger, "North Koreans Launch Rocket Over the Pacific", *The New York Times*, April 4, 2009, at http://www.nytimes.com/2009/04/05/world/asia/05korea.html?_r=0 (Accessed January 10, 2017).
3. "Treaty on Principles Governing the Activities of States in the Exploration and Use of Outer Space, Including the Moon and Other Celestial Bodies (Outer Space Treaty)", NTI, July 13, 2016, at http://www.nti.org/learn/treaties-and-regimes/treaty-principles-governing-activities-states-exploration-and-use-outer-space-including-moon-and-other-celestial-bodies-outer-space-treaty/ (Accessed January 10, 2017).
4. "North Korea Successfully Launches Satellite: Reports", *Space.com*, December 12, 2012, at http://www.space.com/18867-north-korea-rocket-launch-satellite.html (Accessed January 12, 2017).
5. "North Korea Advances 'Satellite' Launch", *BBC News*, February 6, 2016, at http://www.bbc.com/news/world-asia-35511647 (Accessed January 12, 2017).
6. "North Korea fires long-range rocket despite warnings", *BBC News*, February 7, 2016, at http://www.bbc.com/news/world-asia-35515207 (Accessed January 12, 2017).
7. "Treaty on Principles Governing the Activities of States in the Exploration and Use of Outer Space, Including the Moon and Other Celestial Bodies (Outer Space Treaty)", No. 3.
8. "North Korean Satellite 'Tumbling in Orbit' – Reports", RT, February 11, 2016, at https://www.rt.com/news/331810-satellite-tumbling-north-korea/ (Accessed January 12, 2017).
9. "UN Security Council Resolutions on North Korea", Arms Control Association, March 2016, at https://www.armscontrol.org/factsheets/UN-Security-Council-Resolutions-on-North-Korea (Accessed January 12, 2017).
10. Kelsey Davenport, "N. Korea Launch Spurs Talk of New Policy", Arms Control Association, February 2013, available at http://legacy.armscontrol.org/act/2013_01-02/North-Korea-Launch-Spurs-Talk-of-New-Policy%20 (Accessed January 12, 2017).
11. Frederic L. Kirgis, "North Korea's Withdrawal from the Nuclear Nonproliferation Treaty",

American Society of International Law, 8 (2), January 24, 2003, at https://www.asil.org/insights/volume/8/issue/2/north-koreas-withdrawal-nuclear-nonproliferation-treaty (Accessed January 18, 2017).

12. "Treaty on Principles Governing the Activities of States in the Exploration and Use of Outer Space, Including the Moon and Other Celestial Bodies", United Nations Office of Outer Space Affairs, December 19, 1966, at http://www.unoosa.org/oosa/en/ourwork/spacelaw/treaties/outerspacetreaty.html (Accessed January 18, 2017).
13. "North Korea's Nuclear and Missile Programs", International Crisis Group Asia Report, No. 168, June 18, 2009.
14. Samuel Ramani, "The Long History of the Pakistan-North Korea Nexus", *The Diplomat*, August 30, 2016, at http://thediplomat.com/2016/08/the-long-history-of-the-pakistan-north-korea-nexus/ (Accessed January 18, 2017).
15. "Top 10 Largest Economy Countries in Asia 2017", at http://www.worldblaze.in/top-10-largest-economy-countries-in-asia/2/ (Accessed January 18, 2017).
16. James Clay Moltz, "The KSLV I Launch and South Korea's Space Strategy", Council on Foreign Relations, October 2012, at http://www.cfr.org/south-korea/kslv-launch-south-koreas-space-strategy/p29292 (Accessed January 18, 2017).
17. "Treaty on Principles Governing the Activities of States in the Exploration and Use of Outer Space, Including the Moon and Other Celestial Bodies", US Department of Space, at http://www.state.gov/t/isn/5181.htm (Accessed January 27, 2017).
18. Ajey Lele, "East Asia's Space Agenda", *Asian Space Race: Rhetoric or Reality?*, Springer, New Delhi, 2013, p. 73.
19. "U.S. Relations with the Republic of Korea", US Department of State, January 25, 2017, at https://www.state.gov/r/pa/ei/bgn/2800.htm (Accessed January 27, 2017).
20. Ibid.
21. "US-South Korea Relations", Congressional Research Service, October 20, 2016, at https://fas.org/sgp/crs/row/R41481.pdf (Accessed January 27, 2017).
22. "Full-Spectrum Dominance of Outer Space Can Turn Frontier into 'Military Theatre', Build Walls of Suspicion, Breach Global Security", United Nations General Assembly, October 23, 2012, at http://www.un.org/press/en/2012/gadis3464.doc.htm (Accessed January 27, 2017).
23. David R. Sands, "Russia, China React Angrily as U.S. to Put Anti-missile THAAD System in South Korea", *The Washington Times*, July 8, 2016, at http://www.washingtontimes.com/news/2016/jul/8/russia-china-angered-us-anti-missile-system-skorea/ (Accessed January 27, 2017).
24. "Full-Spectrum Dominance of Outer Space Can Turn Frontier into 'Military Theatre', Build Walls of Suspicion, Breach Global Security", No. 22.
25. James Clay Moltz, No. 16.
26. "Law Concerning the National Space Development Agency of Japan (Law No. 50)", June 23, 1969, at http://www.spacelaw.olemiss.edu/library/space/Japan/1969-50%20-%20NASDA.pdf (Accessed January 22, 2017).
27. Manuel Manriquez, "Japan's Space Law Revision: the Next Step Toward Re-Militarization?", NTI, January 1, 2008, at http://www.nti.org/analysis/articles/japans-space-law-revision/ (Accessed January 22, 2017).
28. "Basic Space Law of Japan (Law no. 43 of 2008)", August 27, 2008, at http://stage.tksc.jaxa.jp/spacelaw/country/japan/27A-1.E.pdf (Accessed January 22, 2017).
29. Yasuyoshi Komizo, "Japanese Space Policy – Basic Space Law", Permanent Mission of Japan

to the International Organisations in Vienna, at http://www.unoosa.org/pdf/pres/lsc2009/pres-09.pdf (Accessed January 22, 2017).

30. Setsuko Aoki, "Japan Enters a New Space Age", *The Asia Times*, July 3, 2008, at http://www.atimes.com/atimes/Japan/JG03Dh01.html (Accessed January 22, 2017).

31. John Gittings, "North Korea Fires Missile over Japan", *The Guardian*, September 1, 1999, at https://www.theguardian.com/world/1998/sep/01/northkorea (Accessed January 25, 2017).

32. Manuel Manriquez, No. 27.

33. "North Korea's Nuclear Threats", *The New York Times*, April 16, 2013, at http://www.nytimes.com/interactive/2013/04/12/world/asia/north-korea-questions.html?_r=0 (Accessed January 25, 2017).

34. Kiyoshi Takenaka, "Japan Says Faces Increasing Threats from China, North Korea", *Reuters*, July 9, 2013, at http://www.reuters.com/article/us-japan-defence-idUSBRE96808T20130709 (Accessed January 25, 2017).

35. Richard Spencer, "Chinese Missile Destroys Satellite in Space", *The Telegraph*, January 19, 2007, at http://www.telegraph.co.uk/news/worldnews/1539948/Chinese-missile-destroys-satellite-in-space.html (Accessed January 27, 2017).

36. Saadia M. Pekkanen, "All Eyes on China, But Japan May be the Space Power to Watch", Forbes, May 30, 2015, at http://www.forbes.com/sites/saadiampekkanen/2015/05/30/all-eyes-on-china-but-japan-may-be-the-space-power-to-watch/#30f6b4e14476 (Accessed January 27, 2017).

37. Maeda Sawako, "Transformation of Japanese Space Policy: From the 'Peaceful Use of Space' to 'the Basic Law on Space'", *The Asia Pacific Journal*, 7 (44:1), November 2, 2009, at http://apjjf.org/-Maeda-Sawako/3243/article.html (Accessed January 25, 2017).

38. "Japan's Disarmament and Non-Proliferation Policy: Third Edition", Ministry of Foreign Affairs of Japan, at http://www.mofa.go.jp/policy/un/disarmament/policy/pamph0603.html (Accessed January 25, 2017).

12

Impact of Outer Space Treaty on Few Observer Organisations in COPUOS

Malay Adhikari

Introduction

The Outer Space Treaty (OST) came into force in 1967 against the backdrop of inter-continental ballistic missile (ICBM) testing by two super powers the US and the former USSR – and rapid development of space activities in the 1950s and 1960s. Therefore the United Nations which was also established just after Second World War became very much afraid from the scientific activities going to be performed in outer space by the two superpowers. The outcome is the Outer Space Treaty, 1967 (OST). The main objective of the OST is to promote peaceful uses of outer space. It bans the use of any type of weapons of mass destruction in outer space. However, during the last 50 years, the scenario of countries using outer space has changed. There are now no longer just two superpowers, and several critical activities are taking place in outer space. It appears that using the ICBMs today is a remote concept even to the two oldest superpowers. Moreover space activities have been highly commercialised during the last 50 years. Space programmes previously were shown as status of the countries. This attitude has changed mainly to the direction of commercialisation of space. In addition, there are various other non-government actors. Today, space programmes are shared by state and private actors. The field of space as witnessed major developments last 50 years. But the texts of the OST are so powerful and effective that it is the only treaty in space law that has been ratified by the highest number of countries.

The first part of the paper provides a very brief analysis of the provisions of the OST, analysing how this treaty facilitates the advancement of science and technological experiments in space and promotes peace by banning nuclear weapons in outer space. The second part provides different types of observer organisations in COPUOS. The third part discusses how the provisions of the OST have impact on the observer organisations of COPUOS. The fourth part concludes the paper.

I. ANALYSIS OF THE OST

There are four directions of the OST relevant in the present context. First is how this treaty supports the promotion of science and technology in outer space. Second is the traditional effect of the OST to maintain peace in outer space. Third is promoting international co-operation. It is important to note that the majority of states were non-space-faring when the OST came into existence. Fourth is how this treaty is applicable to international organisations though states are the parties of the treaty. The last one shows why the views of observer organisations in COPUOS about the OST are significant justifying the purpose of this paper.

(i) Promotion of Science & Technology

The title of the OST mentions "activities of states in the exploration and use of outer space". The interpretation of these words cannot be completed without referring to scientific and technological activities in outer space. The following analysis provides more on this issue.

In some parts of the OST, the word 'exploration' is specifically connected with science. The fourth paragraph of the Preamble directly speaks for international co-operation for contribution in scientific aspects of exploration.

The third paragraph of Article I links directly with scientific investigation. It calls for no embargo on scientific investigation in outer space, and encourages states to facilitate this investigation internationally. Therefore it supports international efforts for scientific investigation in outer space.

The second paragraph of Article IV mentions 'scientific research'. It provides promotion of scientific research through the use of military personnel. It further encourages the use of any equipment or facility for such purpose.

Article IX insists on pursuing studies of outer space and conducting its exploration to prevent environmental damage to the earth due to the entry of extraterrestrial matter. It also mentions taking appropriate measures for this purpose. The 'study' and 'measures' both would require considerable input from science and technology. So indirectly this Article promotes the study of environmental science.

A part of Article XI is addressed to international scientific community. It calls for states to update the community about the activities in outer space, thus implying the enrichment of scientific knowledge globally.

Besides these provisions, the other provisions of the OST also have an indirect link with space science and technology.

(ii) Maintaining Peace by Banning Weapons of Mass Destruction in Space

It is already mentioned why the OST was formed. The main purpose of the OST is to maintain peace by banning nuclear weapons or weapons of mass destruction in outer space. The Preamble along with Articles I to XIII state this very purpose directly and indirectly. Even promoting scientific activities in space are considered in the OST only for peaceful purposes.

(iii) Promotion of International Co-operation

The OST promotes international co-operation in the exploration and peaceful use of outer space. The provisions of this treaty facilitate the global dissemination of scientific and technical knowledge required for conducting activities in outer space. This was especially significant in the earlier decades as there were only a few space-faring countries; and there were even nations that were totally unaware about the importance of space programmes. Moreover, for most countries space is out of reach because of the high cost and huge investment involved. This treaty spreads the awareness of about the space programmes to such countries. As per the treaty, the space-faring countries have a legal duty to share the developments of activities in outer space with the rest of the world.

The third paragraph of the Preamble of OST states that outer space activities should be conducted for the benefit of all peoples, their economic or scientific development notwithstanding. Again, Article I repeats the same. Exploration and use of outer space for peaceful purposes is to be pursued by all states without discrimination of any kind. These provisions expand the scope of international co-operation. Articles III and X specifically promote international co-operation.

(iv) Applicability towards International Organisations

Though the states are parties to the OST, but Article VI mentions international organisations as well. It states that the international organisations, including non-governmental organisations, conducting activities in outer space should follow the provisions of the OST and that they must bear the responsibility for compliance.

Article XIII mentions that the provisions of the OST are applicable to international intergovernmental organisations too.

II. DIFFERENT OBSERVER ORGANISATIONS IN COPUOS

There have been 35 observer organisations in COPUOS since from 1962 to 2016. But this paper would like to confine the present analysis within international organisations as observers. Because they are in majority. The minority group consists of bodies like International Space University, Prince Sultan Bin Abdulaziz International Prize for Water (PSIPW) etc. The observers may be divided into three categories: the first consisting of only intergovernmental organisations, the second a mixture of intergovernmental and non-governmental organisations and the third purely non-governmental organisations.

The first category consists of the European Space Organisation (ESA), International Telecommunication Satellite Organisation (ITSO), International Organisation of Space Communications (INTERSPUTNIK), International Mobile Satellite Organisation (IMSO), Committee on Earth Observation Satellites (CEOS), International Institute for Applied Systems Analysis (IIASA), European Organisation for Astronomical Research in the Southern Hemisphere (ESO), Asia-Pacific Space Cooperation Organisation (APSCO) and Inter Islamic Network on Space Sciences & Technology (ISNET).

The second category consists of the Committee on Space Research (COSPAR), International Astronautical Federation (IAF), International Society for Photogrammetry & Remote Sensing (ISPRS) and European Space Policy Institute (ESPI).

The third category consists of the International Law Association (ILA), Association of Space Explorers (ASE), International Astronomical Union

(IAU), International Astronomical Union (IAU), National Space Society (NSS), Space Generation Advisory Council (SGAC), World Space Week Association (WSWA), International Institute of Space Law (IISL), European Telecommunications Satellite Organisation (EUTELSAT-IGO), International Association for the Advancement of Space Safety (IAASS), International Air Transport Association (IATA), Scientific Committee on Solar-Terrestrial Physics (SCOSTEP) and African Association of Remote Sensing of the Environment (AARSE). The number of observers in this category reaches maximum.

Notably, though all are international organisations, but headquarters of the most organisations are situated either in Europe or the US. It implies that the older space-faring countries have a major impact on these organisations' space-related activities. Therefore, these countries have great influence on these organisations' views or interactions with the OST. The impact of the OST on these organisations is directed by the old space-faring states. These states may interpret the provisions of the OST according to their own terms and requirements. So the international organisations originated in Europe or the US may follow the OST accordingly.

Now if the aforementioned four directions of the OST are simultaneously justified, it will be more appropriate to consider observer organisations based in the Asia-Pacific or African region. Particularly, as the countries in these regions had not developed space activities during, and even after, the period of formation of the OST. Even now, there are some countries that depend entirely on international co-operation for pursuing any activities concerned with space. The examples may be Bangladesh and Sri Lanka.

The rationale behind considering four directions together is as follows: the spirit of the words 'international co-operation' as used in the OST should be really tested in a region where space activities have not yet been developed. The question is whether the international organisations formed in such regions not earlier developed in space, could exert a significant sphere of influence globally. It is important to note that the spirit and provisions of the OST do not allow supremacy or dominance of any international organisation over new or smaller ones.

In the aforementioned background, of all the observer organisations, some have permanent status like APSCO, ISNET and AARSE, which are considered here.

III. IMPACT OF THE OST ON OBSERVER ORGANISATIONS OF COPUOS

(A) APSCO

APSCO was established in 2005 through a convention. It is an inter-governmental international organisation as per Article 3 of the APSCO Convention. So Article XIII of the OST is fully applicable to it. The member countries are Bangladesh, China, Iran, Mongolia, Pakistan, Peru, Thailand and Turkey. Of all these member nations, only China, Pakistan and Iran have significant space programmes as of now. It shows that APSCO being an international organisation has been established in a region that is underdeveloped in space activities. In this sense, the study of impact of the OST on APSCO is more appropriate.

(i) Promoting Peace

The first paragraph of Preamble of the Convention mentions peaceful use of space technology. The second paragraph stresses on multilateral co-operation in the Asia-Pacific region for peaceful applications of space science and technology. The last paragraph of the Preamble refers to the 'principles of peaceful uses of outer space, mutual benefits and complementariness, equal consultations and development'. These are similar to the spirit and provisions of the OST. Articles 4 (1, 5) of the Convention provides for peaceful application of space science and technology as the objective of APSCO.

(ii) Promoting International Co-operation

(a) APSCO Convention

The Convention Preamble mentions words like 'multilateral co-operation' or 'regional multilateral co operation' which convey its international spirit. At the same time, they also imply that a region with low awareness of space programmes first requires regional co-operation. Article 4(5) ("Objectives") of the Convention aims to contribute to the peaceful uses of outer space in the international co-operative activities in space technology and its applications. The objective is also to provide collaborative space programmes through co-operation in the Asia-Pacific region.

Article 9(6) provides that any international organisation may enjoy the observer status with APSCO if it is granted by its Council. Not only that, the states outside the Asia-Pacific region may join as Associate Members satisfying the prescribed conditions. Article 24 provides co-operation with other entities. The sub-section (2) states very clearly the ambit of co-operation. The organisation could make co-operative partnership with the non-member states and different international organisations and institutions subject to the permission of APSCO Council. So the scope becomes very wide.

Thus, although APSCO is centred in the Asia-Pacific region but it is not limited to the region. In other words, just like international observer organisations centred in Western countries have members from the Asia-Pacific region, APSCO also is open to countries from other regions. Article 9(6) states that any member state of United Nations (UN) or any international organisation may be observer in the Council's meetings. Another provision in Article 9(7) provides that any member state of UN outside the Asia-Pacific region may be Associate Member without any voting right in the Council's meeting. Both these provisions are with the permission from APSCO Council. It facilitates co-operation not only among countries in the Asia-Pacific region but also among other international organisations.

(b) Introductory Document of APSCO

Besides the Convention, the preface of introductory document of APSCO declares why international co-operation in space programmes is necessary especially for countries lacking in space technology and economic resources. The simple reason is when any country does not have any space programme, they have to share the same from the countries that are enriched in space activities. And they have to follow the terms and conditions as set by the countries developed in space.

(iii) Promoting Science & Technology

(a) APSCO Convention

The promotion of science and technology is mentioned in many terms directly and indirectly in the Convention text. The third paragraph of the Preamble directly states about the development of space science and technology. The objectives under Article 4(2, 3) provide for the promotion of science and technology. The basic activities (under Article 7) and optional activities (under Article 8) of APSCO have many provisions for the development of space science and technology along with education and training activities.

(b) Beijing Declaration

The Beijing Declaration 2015 is another significant document of this organisation. The fourth paragraph of its Preamble mentions the peaceful uses of outer space through Asia-Pacific multilateral co-operation. The actions are to be implemented in order to develop capacity related to space science and technology and its education and training; and co-operation through improved sharing service capability (sharing of data etc.). Improving quick response capability and information inter-connection are aimed at better co-operation amongst Asia-Pacific countries. The text of this Declaration ends with repetition of words like 'peaceful utilisation', 'mutual benefit' for APSCO and its member states. These words are used in the OST too.

(B) ISNET

ISNET was established by nine Organisation of Islamic Co-operation (OIC) countries in 1987, and was granted the observer status in COPUOS in 2013. Presently, it has 16 member countries, namely Bangladesh, Iraq, Indonesia, Morocco, Niger, Pakistan, Saudi Arabia, Tunisia, Turkey, Syria, Iran, Sudan, Azerbaijan, Senegal, Egypt and Jordan. Out of these, there are hardly four countries that have space programmes. ISNET promotes space science and technology according to the provisions of the OST.

One of the main objectives of ISNET is 'to collaborate and co-operate with OIC member countries in the peaceful uses of outer space'. It fully supports the provisions and spirit of the OST. The rest of the objectives are drafted for developing and exchanging space science, technology and their applications.

Further, ISNET requires member countries 'to share their experiences, research studies and developments in space sciences and applications in their respective countries...'. It reflects the principle of mutual assistance as mentioned in Article IX of the OST. The third paragraph of the OST Preamble states that exploration and use of outer space is for the benefit of all peoples. A similar tone is observed in the ISNET objective, 'exploiting the potential of the outer space for the development and benefit of these countries.'

Recently, ISNET, which is purely dedicated to the development of space science and technology, has started co-sponsoring the Second International Conference on Space (ICS-2016). The ICS-2016 has major sessions on space science, technology and applications, but only a smaller session on space law, policy and regulatory issues. Space law may be called the tail part of ICS-2016. It seems to be the fact that ISNET feels later the importance of space law in developing space science and technology. But space law cannot be discussed or analysed without OST in ICS-2016. Therefore indirectly this organisation feels the importance of the OST in promoting their activities.

(C) AARSE

AARSE is a non-governmental organization organisation with permanent observer status granted since 2014. It was incorporated in 2008.

Some of the objectives of AARSE are as follows: (1) Mutual sharing of benefits between international space science, technology and application programmes and that of between African countries; (2) promotion of greater co-operation and coordination among African countries; (3) assistance of regional and international user communities through dissemination of scientific, technical, policy and programme information of AARSE; and (4) exchange of views and ideas concerned with earth observation system and geo-information science for the betterment of Africa.

The first objective reflects the principle of mutual sharing as described in Article IX of the OST. The second and third objectives support the principle of international co-operation as observed in the OST. The last one talks about the betterment of Africa, similar to the phrase 'benefit of all people' in the OST. Lastly, it is obvious to say that the objective of AARSE is the promotion of science and technology of remote sensing internationally. Therefore, AARSE follows the spirit of the OST.

IV. CONCLUSION

The OST came into existence in 1967. The observer organisations in COPUOS analysed above are very young compared to the OST, which is normal for any space-related international organisation originating in the Asia-Pacific or Africa. In a majority of the cases, space programme initiatives came about only after the adoption of the OST. Therefore, it is not possible to check the impact of the OST on these organisations. It may be said that it would require sufficient time to observe the impact of the OST in the activities of such observer organisations.

But there is some sort of general conclusion for these organisations. There are no derivations from the provisions of the OST in the rules and regulations or objectives of these observer organisations in COPUOS. Even it may be forecasted that the same trend will continue in near future. The reasons are obvious and natural. The OST was founded on considering great prospect of humankind in outer space. It is in the very first line of its Preamble. The treaty speaks for humanity in different modes and tones through its texts. Additionally, the backbone of the treaty is according to the purposes and principles of the UN Charter, and the activities in exploration and use of outer space in accordance with international law. In all these senses, it is very difficult to escape for any international organisation in general from the purview of the OST. These international organisations as observers in COPUOS are originated in region that is underdeveloped in space programmes. Therefore they have enjoyed great support and strength through the OST to expand their organisational activities. This treaty is a legal shield for them to prevent any dominance by older space-faring states or international organisation originated in Western countries.

Moreover in the era of commercialisation and privatisation of space activities, the OST facilitates through its provision of international co-operation to spread space commerce from underdeveloped region to developed one. So international organisations like APSCO could very well utilise the OST to promote international space commerce from Asia-Pacific region.

Therefore all in all alike international organisations should welcome the grand fifty years of OST.

REFERENCES

1. The Outer Space Treaty, 1967.
2. Committee on the Peaceful Uses of Outer Space: Observer Organisations, http://www.unoosa.org/oosa/en/ourwork/copuos/members/copuos-observers.html
3. The Asia-Pacific Space Co-operation Organisation (APSCO), http://www.apsco.int/
4. Inter-Islamic Network on Space Sciences & Technology (ISNET), http://isnet.org.pk/
5. 2nd International Conference on Space 2016 (ICS-2016), http://suparco.gov.pk/webroot/pages/ICS-2016.asp
6. African Association of Remote Sensing of the Environment (AARSE), http://www.africanremotesensing.org/

SECTION III

GOVERNANCE AND PROSPECTS

13

Evolution of Policy and Law for International Space Governance

Eligar Sadeh

This chapter examines the evolution of policy and law directed at international space governance since the rise of the space age with a focus on the challenges for the global space community. The more states, non-governmental actors and commercial players depend on space, the more policies, such as norms, best practices, rules and compliance mechanisms, and laws, like the Outer Space Treaty (OST) regime, are essential to maximise benefits related to uses of space. Realisation of benefits is linked to effective international governance that secures the space domain for peaceful and sustainable uses; protects space assets from the risks of orbital debris and irresponsible behaviour in the space domain; and derives value from space assets for security, economic, civil and environmental applications and value-added services.[1] In this context, the concepts of global commons and strategic stability are applicable.[2]

Global commons relates to governance in accordance with the OST regime on the basis of voluntary actions, self-restraint and self-regulation. One challenge to space governance highlighted with global commons is that incentives to tradeoff national interests and gains in return for collective action and international benefits are generally not present.[3] Collective action is difficult to sustain with competitive security relationships among major spacefaring states, as well as with the increasing number and variety of space actors worldwide, many of which are commercial and non-state actors. All this makes space not only more competitive, but more congested. Issues of space governance are discussed through the framework of a global commons with regard to the legal norms and principles of the OST regime,[4] and the cases of policy coordination related to mitigating orbital debris and to optimally making use of Earth observation data.

Strategic stability is governance anchored on shared strategic goals, namely the security, commercial, civil and environmental benefits gained from space. The challenge for stability lies in advancing credible strategic assurance. Although strategic assurance includes protection and deterrence related to space assets, it also concerns sustainable uses of space, freedom of access and use of space and freedom from threats to space assets.[5] Without assurance, sustainable use of space is undermined and space assets operate in a contested milieu. Analysed within the scope of strategic stability are lawful means of space protection, deterrence policies to protect space assets and policy approaches to global engagement focusing on capacity building, confidence building measures and codes of conduct in the space domain.

Global Commons: Challenge of Collective Action

The space domain is considered a global commons in that the domain lies beyond the sovereign jurisdiction of states, is governed by international law and is available for all actors to access and use. This includes free space itself, orbital paths around the Earth and celestial bodies. A commons is in joint supply and use, and nationally non-appropriable. Joint supply and use signify equal potential availability to the commons by all states. Non-appropriability specifies that states cannot extend their jurisdiction and sovereignty to the commons. It is impossible to exclude states from sharing in the benefits of the commons or from suffering the consequences caused by damage to the commons. Joint supply and non-appropriability constitute free access and free use, and in the case of the commons of space, freedom of action in space.[6]

A commons with no governance, and thus unregulated by law, results in a "tragedy of the commons".[7] This situation is rooted in rational, self-interested behaviour regarding the commons. The tragedy is a function of damage to the commons caused by free access and free use, like the proliferation of orbital debris, the possibility of interference and attacks on space assets and harmful contamination of space and celestial bodies. The commons of space posits a collective action problem: how to formulate and implement space governance to address these tragedies, and optimally restrain and regulate at some level free access and free use. Of importance to states is the goal of creating a just and sustainable international order that can foster collective action to confront common challenges.[8] Such an order is plausible through global governance that is rooted in a rules-based international system, which advances self-interest by serving mutual interests. Rulemaking by institutional arrangements, which range from voluntary policies to formal and legally binding treaties and agreements, frames the scope of governance.

OST Regime

Space is characterised as a commons based on freedom of action there. This points to the collective action problem discussed above. The OST regime supports the free use of and free access to the space environment for peaceful purposes that include civil, commercial and military uses. Although this regime provides the basic legal framework for governing space as a global commons and serves as a basis for addressing challenges to collective action, the regime falls short in realising collective action to space governance. The regime does not provide (a) detailed rules and an authoritative process for deciding what types of space activities are inconsistent with its key principles; (b) when the use of space might damage common interests; and (c) benefit sharing from space activities.[9] In fact, the OST regime tends to reflect the tragedy of the commons since self-interested users, the spacefaring states, exploit relative gains from using space with little regard to negative effects on other users and on the space domain itself.

Averting a tragedy of the commons in space implies that these negative effects are considered. The options for states are to either establish alegally binding central governing authority to make rules, verify compliance and respond to violations, or apply customary law and policies of self-restraint and self-regulation to ensure sustainability of the domain. Given that the former option does not exist along with the shortcomings of the OST regime, the latter serves as the viable way forward to attain collective action. In addition to customary law, policies addressing international norms, codes of conduct on behaviour and use, transparency and confidence building measures and diplomacy play key roles for governance. These means, however, depend on actors placing a high value on collective benefits in relation to self-interested ones, and thus, a high degree of self restraint and self-regulation to avoid irresponsible behaviour in the space domain.

Despite prospects at collective action, the advent of congested space and irresponsible uses of space impose risks that space actors will cause problems for each other, whether purposeful or not. One case in this respect deals with minimising and mitigating orbital debris, which damages space assets and undermines sustainable uses of orbital space. Collective action in-and-of-itself, which is supported by the OST regime, also implies international cooperation to address problems. The case of cooperation and coordinating the use of Earth observation missions and data is illustrative here. In both cases, spacefaring actors prefer modes of self-governance and voluntary international policy coordination to maintain political and operational flexibility.

Orbital Debris

There currently exists 17,817 orbital debris objects large enough to be tracked (as of 4 October 2016, US Space Surveillance Network), and an unknown number of smaller debris objects that are either invisible to sensors or that cannot be tracked on a regular basis, but nonetheless a risk for space assets.[10] Orbital debris includes fragmentation debris (satellite break-up debris and anomalous debris events), spacecraft, mission-related debris (objects dispensed, separated, and released as part of a planned mission) and rocket bodies in orbit around Earth. With cascading effects (debris colliding with debris in a cycle of debris creation), some orbits will become more dangerous and others may no longer be useable in the future. States recognise the problem and encourage voluntary debris mitigation guidelines. These guidelines represent evolution in policy for successful collective action.

With no collective action, where no active mitigation measures are implemented, fragmentation debris together with cascading effects shifts linear debris growth patterns to exponential ones. The debris problem is also exacerbated by the proliferation of thousands of additional debris fragments that threaten space assets to this day due to irresponsible space behaviour and orbital conjunctions. The Chinese anti-satellite (ASAT) weapon test conducted in January 2007, which destroyed a Chinese satellite, exemplifies the former, and the February 2009 collision between an operational Iridium satellite and a dysfunctional Russian Cosmos communications satellite demonstrates the latter.

The majority of operational and active satellites are vulnerable to orbital debris and impacts. The failure to prevent debris proliferation in low Earth orbit (LEO) results in an issue of restricted uses of the more commonly used orbital paths and orbital inclinations. Hence, mitigation is needed in LEO. To add, the debris issue in geostationary orbit (GEO) is potentially serious and costly due to the relative permanency of orbit (no passive debris removal through orbital decay), the narrow orbital band that exists at GEO and the high economic values of GEO orbital slots with lucrative footprints on Earth for telecommunication services.

The functional necessity of addressing the orbital debris issue advances collective action. This is no better exemplified than by the Inter-Agency Space Debris Coordination Committee (IADC). The IADC includes the European Space Agency (ESA) and national space agencies of Italy, France, China, Canada, Germany, India, Japan, South Korea, Russia, Ukraine, the US and UK.[11] The approach taken by the IADC encompasses voluntary policy actions that states take to reduce debris to the establishment of guidelines and standards to govern launch vehicles and their payloads. Principal technical approaches to reduce debris encompass passivation, parking orbits and hardware designs.

Spacefaring states at the national level also address debris mitigation. Beginning in the 1990s, the US, ESA and spacefaring actors developed national guidelines to reduce the production of debris during launch and with on-orbit operations, move GEO satellites into parking orbits at the end of their service life and put defunct LEO satellites into decay orbits. Following such best practices involves additional costs, complicates operations and shortens the useful life of satellites. Therefore, national requirements, compliance and enforcement levels vary from state to state.

To harmonise and strengthen national practices, the United Nations Committee on Peaceful Uses of Outer Space (UNCOPUOS) developed international guidelines for orbital debris mitigation that were adopted in 2007 and endorsed by the United Nations General Assembly (UNGA) in 2008.[12] The guidelines, *Space Debris Mitigation Guidelines of the Committee on the Peaceful Uses of Outer Space*, are an achievement for space governance, given the pledges not to deliberately create debris. A 2009 UNCOPUOS report concluded, "Implementation of the voluntary guidelines for the mitigation of space debris at the national level would increase mutual understanding on acceptable activities in space, thus enhancing stability in space and decreasing the likelihood of friction and conflict."[13]

Even though these guidelines for the mitigation of space debris are voluntary, they are reflected in existing best technical practices.[14] One issue though is the vague language of the guidelines, which allows each space user to decide how many design and operational changes are reasonable to limit debris production, minimise break-up potential, reduce the probability of accidental collision and avoid intentional destruction in ways that produce long-lived debris. Since these measures are voluntary, the guidelines are not completely effective although there are improvements in compliance. For example, only 11 of 21 GEO spacecraft that ended their service life in 2009 (one year after UN endorsement of the guidelines) were disposed of properly; yet, by 2014, 13 of 18 GEO spacecraft at end of service complied with guidelines.[15]

Clearly, the formulation of international guidelines for orbital debris is an important step to more formalised collective action on the debris issue. This is important as the OST regime deals with space objects that are registered as a legal remedy for liability issues. As such, questions remain as to legal definitions of debris and whose debris is causing harm, especially if that harm is in the space environment and under the fault-based liability regime per the *Convention on Registration of Objects Launched into Outer Space*.

Collective action progress with regard to debris mitigation illustrated by the IADC, national efforts and the UN guidelines are noteworthy, though shortcomings exist in the long-term sustainability in managing space on the basis of a global commons. These shortcomings are present since policy preferences for voluntary actions, self-restraint and self-regulation lead to outcomes of self-interest versus collective action. Collective action is impeded when spacefaring states resist actions for the long-term sustainable management of space as a global commons, especially if they believe that access to and use of space is used for gains relative to potential competitors.[16] Space as a global commons is enabled when states not only assert rights to use space without interference from others, but also acknowledge that other users retain similar rights, and that all rights in space confer corresponding responsibilities to ensure sustainable uses of space.[17]

Earth Observations

International cooperation pertaining to Earth observations by satellites directed at assessing global environmental change is represented by a collective action milieu. The goal of this collaboration is to advance scientific knowledge of the Earth's environment to understand and predict human-induced and natural global environmental change phenomena. One of the crucial factors in this case is the ability of transnational networks of Earth system scientists to work together in analysing global change data and to translate those analyses into policy relevant actions. This involves both coordinating missions and addressing data policy issues dealing with conditions and access to data, data pricing, periods of exclusive data use and data archiving. International cooperation aims to meet scientific and operational needs and satisfy data access and data exchange requirements for all parties as effectively as possible.

Political considerations concerned with data policy, national sovereignty and national security issues influence collective action in the area of Earth observations.[18] The existence of disparate and incompatible data access policies among various satellite types and programmes is reinforced in the retention of data by its producers, the requirement of licences to use data and the pricing of data above marginal costs of fulfilling user requirements. Harmonising policies over these issues is a collective action challenge to surmount.[19]

The Committee on Earth Observation Satellites (CEOS) plays a central role in advancing harmonisation. The goals of CEOS are to optimise the benefits of Earth observations through cooperation of its members in mission planning, and in developing compatible data products, formats, services, applications and policies; aid both its members and the international user community through international coordination of Earth observation activities; and exchange policy and technical information to encourage complementarity and compatibility among Earth observation systems.[20] CEOS data exchange principles exist for global environmental change research use and for operational public benefit use with the agreement to make data available to each member in these user categories with no period of exclusive use and on a non-discriminatory basis. There is a commitment to provide data at the lowest possible cost to researchers and to harmonise and preserve data needed for long-term global change research and monitoring.

To further advance coordination among national Earth observing systems, the Group of Eight leading industrialised countries (G8) during a 2003 meeting supported additional collaboration. The G8 recommendation led to the establishment of the Group on Earth Observations in 2005. Today, this group includes 101 governments and the European Commission, as well as additional inter-governmental, international and regional organisations. Even though participation and funding are voluntary, the Group on Earth Observations advances collaboration in systems architecture and interoperability, data management and capacity building associated with Earth observing systems.[21]

Even though there exists collective action to avoid duplication, coordinate coverage and take steps to synchronize operations, as advanced by both CEOS and Group on Earth Observations, the national sovereignty of the natural resources being observed – air quality and land use, for example – conflict with coordination. Since remote sensing data undercuts the ability of the state to control both the creation and the application of knowledge, there are concerns with sovereignty and national security. One problem in using space-derived Earth observations deals with platforms owned by one state to assess the natural resources of other states, particularly when the resources are of economic value.[22] With regard to civil systems (excludes military and commercial systems), the UN principles of remote sensing allow observations of other countries within the framework of cooperation.[23] The principles require that space-collected observations of sovereign resources of the sensed country be provided to the sensed country.

> As soon as the primary data and the processed data concerning the territory under its jurisdiction are produced, the sensed State shall have access to them on a non-discriminatory basis and on reasonable cost terms. The sensed State shall also have access to the available analysed information concerning the territory under its jurisdiction in the possession of any State participating in remote sensing activities on the same basis and terms, particular regard being given to the needs and interests of the developing countries.[24]

Sovereignty issues are also a concern in the case of commercial remote sensing systems. Proliferation of high-resolution imagery with intelligence value posits national security repercussions for several reasons. First, increased certainty of an adversary's capabilities negates the foundation for deterrence. Second, the possibility exists of misinterpretation and international deception leading to shifts in balances of power and conflict. And third, asymmetrical access to satellite imagery and processing capabilities provide substantial advantages for some states over others – developed states over developing ones – with destabilising influences on the international system.

An important implication of orbital debris discussed earlier and Earth observations is one of collective action to establish acceptable constraints on state sovereignty to limit freedom of action in and use of the space domain. To this end, international space law and other international agreements that limit military uses of space as explained next put forward one set of constraints, and policy-focused voluntary codes of conduct on behaviour in space discussed herein as well advances another set.

Strategic Stability: Challenge of Strategic Assurance

Strategic stability relates to collective action on the basis of shared strategic goals. The challenge for space governance within this concept lies in credible strategic assurance. A space assurance strategy depends on several elements encompassing lawful means of space protection, deterrence policies to protect space assets and policy-focused approaches to global engagement.[25]

Lawful Means of Space Protection

Realising the protection of space assets as a shared strategic goal begins with lawful means of space protection. The existing system of treaties, customary law, the laws of armed conflict and other legal principles restrict conflict and mitigate threats to space assets. Under treaty and customary law, the right to respond to attacks against space systems and to perform deterrence or protection activities is limited. The use of force is allowed only in cases of self-defence or in accord with authorisation of the UN Security Council to maintain international peace and security.[26]

Space warfare is also constrained by the laws of armed conflict. These laws establish boundaries on the use of force during armed conflicts through application of principles and rules. The principles and rules combine elements of treaty and customary law at the international and national levels. As it concerns space warfare, this body of law sets limits on when, and to what degree, force is used for targeting. Also, self-defence acts that seek to, or actually, damage the space environment are unlawful under the *Environmental Modification Convention* and the OST regime. Obligations to avoid and minimise the creation of orbital debris and the duty to avoid the harmful contamination of space place legal constraints on actions to destroy or damage any space system.[27] International law places limits on military force in space as one key facet of strategic stability. These laws also play a role in maintaining space as a commons.

- *Limited Test Ban Treaty* and *Comprehensive Test Ban Treaty* (supplanted the *Limited Test Ban Treaty*), which prohibit the conduct of nuclear weapons tests in outer space. Of the states with nuclear weapons, the *Comprehensive Test Ban Treaty* is not ratified by the US, China, India, Pakistan, Israel and North Korea. Nevertheless, the treaty with adherence by most other states establishes a norm to follow even for non-ratifying states. In general, multilateral treaties and customs adopted by a large number of subjects of international law (states) are considered universal.
- *Outer Space Treaty* regime, which prohibits the deployment of weapons of mass destruction in space and the stationing of military bases in space or on celestial bodies, and calls for "peaceful uses" of space that is understood as no aggressive uses of space that harm or interfere with another state's access and use of space. The OST regime also prohibits harmful contamination of the space environment.
- *Anti-Ballistic Missile (ABM) Treaty* between the US and Russia, which

was viewed as preventing a weaponisation of space during the Cold War since it prohibited the deployment of space-based ABM systems that do include most types of kinetic-kill and kinetic-energy space weapons. Even though the US withdrew from this treaty in 2002 rendering it null and void, it was effective in advancing an international norm of self-restraint on the deployment of space weapons.

- *Convention on Registration of Objects Launched into Outer Space*, which requires states to register objects launched into space with the UN. This obligation helps to enable space situational awareness and supports the view that such awareness is a shared responsibility to the extent possible without harming national security.
- *Environmental Modification Convention*, which prohibits military use of environmental modification techniques in space.
- *Moon Agreement*, which sought to de-militarise the Moon and celestial bodies. The *Moon Agreement*, however, holds little to no legal validity since no spacefaring powers ratified it to date. Nevertheless, the *Moon Agreement* declares the Moon the "Common Heritage of Mankind". Common Heritage of Mankind differs from the "Province of all Mankind" of the OST regime (i.e., the commons idea discussed earlier in this chapter based on free access and use of space) in that it establishes the natural resources of the Moon as a common property resource for all mankind. If this is accepted, the *Moon Agreement* requires that lunar resources, once exploitation commences, be shared equitably through an international arrangement, such as an international regime. This arrangement supports deterrence (by "entanglement", as discussed below) aimed at peaceful uses of the Moon and advances collective action on the basis of benefit sharing. Of note as well for benefit sharing is that the Common Heritage of Mankind principle is part of the UN *Law of the Sea* as it applies to the exploitation of the deep seabed and the associated regime of governance with the International Seabed Authority.

Deterrence to Protect Space Assets

One way to prevent threats to space assets is to persuade potential aggressors that any benefits from interference are outweighed by expected costs. This is the overall basis for deterrence. The concept of deterrence is applied here to think about how to protect space assets from threats as a shared strategic goal. Deterrence on the basis of international norms and entanglement are useful in this regard.[28] International norms include treaty law and customary law, arms control treaties, test bans, weapons moratoria both formal and informal, confidence building measures and "rules of the road" to govern conduct. Entanglement is a result of interdependence. The question of concern for deterrence is whether norms and entanglement advance deterrent effects that are shared and mutual.

The OST regime establishes a universal set of international norms that are based on treaties and customary law. In relation to deterrence, the regime bans the stationing of nuclear, but not conventional, weapons in orbit and bans military activities on the lunar surface, stating that the Moon and other celestial bodies must be used for peaceful purposes. This represents a shared notion of deterrence based on self-restraint.

Arms control agreements also curtail aggressive actions in space. For example, the key space powers since the rise the space age, namely Russia and the US, tolerated each other's use of space in stabilising ways, such as in the case of space-based surveillance for arms control verification purposes and intelligence gathering. Further, both powers practised reciprocal restraint regarding activities that destabilise the sustainable uses of space, like interference or attacks on space-based surveillance systems. The provisions of the strategic and intermediate range nuclear arms limitation agreements ban interference with national technical means of verification that is enabled by surveillance satellites; both Russia and the US extended the non-interference ban to the entire military space constellations of the other. This engendered a level of stability and predictability to the strategic balance in space. In addition, neither space power pursued ongoing dedicated and operational ASAT options, nor did they place conventional weapons in space on a permanent basis for attacks on space assets. All this demonstrates mutual self-restraint even though both engaged in ASAT work and both were capable of latent retaliatory options if their satellites were attacked.

Verifiable test bans are also effective, as the *Partial Test Ban Treaty* and the *Comprehensive Test Ban Treaty* exhibit. Arms control agreements that verifiably limit testing strengthen deterrence by decreasing an adversary's confidence of success, enhance warning of a change in the strategic environment and prevent an ASAT arms race. Further, an adversary is unlikely to launch a pre-emptive attack with weapons not tested under realistic conditions, and for policy and operational reasons, kinetic-energy ASATs make little sense given orbital debris in such actions. The preference for deterrence tends to be on diplomacy to sway a space attacker. This policy approach is generally viewed as preferable to the use of offensive counterspace options.[29]

Notwithstanding norms, law and self-restraint with ASAT deployment and use, space powers remain concerned about attacks on their space assets.[30] To draw global attention to this potentially destabilising issue, Russia and China became vocal proponents for negotiating on "Prevention of an Arms Race in Outer Space" (PAROS) in the Conference on Disarmament (CD). This agenda item gained near universal support in annual UNGA resolutions, but the US consistently objected on the basis that it is neither possible to define the nature of a space-based weapon nor plausible to develop a verifiable agreement for banning space-based weapons and terrestrial-based ASATs.[31]

In 2008, Russia and China introduced a draft treaty – *Treaty on the Prevention of the Placement of Weapons in Outer Space, the Threat or Use of Force against Outer Space Objects* (PPWT) – which extends the OST's ban on weapons of mass destruction in space to prohibit placing all types of orbiting weapons, including all types of force against space assets.[32] The draft treaty also outlaws deployment of space-based missile defence interceptors, but not debris-generating ASAT tests or the proliferation of ASAT capabilities. Given US objections to PAROS/PPWT and preferences for non-legally binding transparency and confidence-building policies and measures,[33] general concerns with other arms controls issues and issues regarding enforcement and compliance of any potential agreement, the CD is at a standstill. Moreover, the CD is hobbled by the fact that consensus is required, even for procedural matters, and by the linkages among nuclear disarmament, space security and conventional disarmament issues.[34]

Despite CD challenges and disagreements among space powers on specific legal modalities, informal international norms emerging from the CD play a role in fostering shared strategic goals; in this case, one of self-restraint in the use of space weapons, and more broadly, non-legally binding (policy) approaches to collective action for space governance. For instance, while some states supported the commencement of negotiations on PPWT, others voiced reservations that a new legally binding instrument needed to be comprehensive, precise and verifiable. As such, states discussed the merits of confidence-building measures, space debris and political ways to restrain development and testing of destructive anti-satellite weapons.[35]

Deterrence by Entanglement

Deterrence by "entanglement" is the notion that state actors will be deterred from attacking others because of interdependence. The *Moon Agreement* mentioned above supports this type of deterrence given the provisions to equitably share lunar resources. More generally, the degree of globalised interdependence that characterises the modern world is without precedent and ties together spacefaring states in a system of international trade and finance. Satellites are one vital communication node in this trade and finance system.

Any threats or breakdowns in the system of trade and finance are not easily repaired. The destruction of satellite communications destroys global wealth. Reconstruction of the financial system without space assets to restore confidence in reliable trade and financial transactions is a formidable and time-consuming task. Even an attack on a small proportion of the commercial satellite infrastructure brings about consequences for the wealth of all globalised economies. It is difficult to envision any national gains by interference or an attack on these space assets that off-set potential economic losses.

Entanglement extends beyond trade and financial transactions that rely on satellite communications to the various applications of position, navigation and timing (PNT) satellite data.[36] The US, for example, ended the encoding of PNT data in 2000, which degraded the signal provided by the US Global Positioning System (GPS) constellation that was originally built for military purposes. Since then, the precise GPS signal is available globally as a public utility. PNT data is now built into navigation, electrical and transportation grids worldwide, among a vast number of other systems and devices, creating a degree of technological dependence and entanglement. Further, other PNT systems exist among other space powers that include Russia, Europe, China, Japan and India. The PNT case demonstrates deterrence by entanglement; when a system proliferates globally for civil and commercial uses, attempts to deny functions result in global repercussions.

Telecommunication satellite services also highlight deterrence by entanglement. Communication systems originally built for civilian purposes followed by commercial uses now carry a majority of telecommunications bandwidth for military uses. Hostile action to disrupt military communications over commercial satellite systems draws into the crisis numerous other governments whose own military, civilian and commercial traffic is carried by the same satellites. Because the use of commercial satellite-based transponders is market-based and constantly shifting, an aggressor's planning is complicated by the inability to know or effectively predict which other friendly, neutral or potentially adversarial states are affected at any given moment by interference with a particular commercial satellite. Further, global markets for satellite services implies that broad economic consequences will take place with any attack on commercial satellites. This web of mutual dependence and shared consequence acts as a deterrent on threats to these space assets.[37]

An additional dimension of entanglement is tied to international cooperation associated with multinational operations. This is an important component of an effective global engagement strategy to assure access to space capabilities for a state, and for its allies and partners. International cooperation complicates adversary plans and intentions, and creates more stakeholders in the orderly use of the space environment. Deterrence is reinforced if an adversary contends with both the national response of others and international responses. In other words, multinational engagement supports deterrence by denying national benefits of an attack.

Global Engagement

Protection of the space domain is rooted in reconciling national interests with collective action on the basis of international governance. Diplomacy and international engagement help to realise this end. Customary and treaty-based restrictions of international law afford all members of the global space community credible confidence for assured access to space. The system of treaties, conventions and agreements help regularise space activities, and help protect the capabilities of the systems placed on-orbit. The positive attributes of this system of law, as well as challenges, were discussed earlier. Global engagement addresses policy-oriented ways to augment existing space law and norms, including capacity building, confidence building measures and codes of conduct.

In the area of capacity building, where much of the work of UNCOPUOS is focused, information sharing and education are important. Of note is the work of the UN Programme on Space Application that is aimed at building capacity through international workshops, training courses and pilot projects on issues, like satellite navigation systems. UNCOPUOS also oversees implementation of the recommendations emanating from the UNISPACE international conferences (UN Conferences on the Exploration and Peaceful Uses of Outer Space) with the goal of identifying and taking actions designed "to maximise opportunities for human development through the use of space science and technology, and their applications".[38] Similarly, UNCOPUOS follows the UN Platform for Space-based Information for Disaster Management (UN-SPIDER). This programme, which started in 2006, seeks "to provide universal access to all countries and all relevant international and regional organisations to all types of space-based information and services relevant to disaster management to support the full disaster management cycle".[39]

It is clear that such activities are necessary for ensuring the safety and security of all space assets. Newcomers to the arena require assistance not only to most efficiently benefit from the use of space, but also to avoid harmful impact on others. The earlier discussion with regard to deterrence by entanglement demonstrates that as space powers work to build more capable systems and share the benefits with others, such as PNT services, all users are incentivised to protect the systems and to respond against threats to those systems to avoid harmful impacts. The adoption of best practices is required by all spacefaring states for any notion of effective governance as well. One recent success related to this was the development of a set of voluntary guidelines for space debris mitigation examined above.

Confidence building measures are essential to global engagement and the international governance of space. Such measures advance opportunities for transparency between potential adversaries, and improve prospects for dialogue that prevent any future dispute from evolving into armed conflict. Measures involve data sharing, business investments, education and information campaigns at global, national and local levels. For some states, cooperative steps to improve strategic stability in space that eschew binding legal limits in favour of dialogue and confidence-building measures are preferable. This was evident in the case of the CD and the US. Confidence building measures elevate mutual reassurance as a focus for strategic dialogue among spacefaring powers. Transparency measures are also an essential ingredient – to test intentions and to dispel misperceptions that generate unwarranted suspicions and fears of attack.

The development of norms through codes of conduct for the use of space lay the foundation for more robust efforts to mitigate threats to space assets and to avoid conflict. A code of conduct entails a body of voluntary rules for best practices, procedures and behaviour in space activities. The European Union (EU) formulated a voluntary draft code in 2008 (current draft version is from 2014) – *International Code of Conduct for Outer Space Activities* – to promote responsible uses of the space commons.[40] Such a code of conduct helps to maintain the long-term sustainability, safety, stability and security of space by establishing guidelines for the responsible use of space. Albeit the code reiterates principles that spacefaring states endorsed in the OST regime and through the adoption of best practices, it does not add greater clarity nor put forward new mechanisms to decide how principles and practices are applied. In addition, multilateral adoption of the code remains a challenge. Despite multilateral open-ended consultations on the code under UN auspices, UNGA resolutions in 2013 that support the code and UN multilateral negotiations on the code in 2015, the US does not support the specific EU code. Nonetheless, the US (under the Obama Administration) was supportive of the process with the desire to play a direct role in formulation.[41]

Conclusion

Collective action and shared strategic goals among spacefaring states are inextricably tied to a framework of international space governance of treaties and agreements, norms, intergovernmental organisations for rule-making, monitoring capabilities and joint decision-making. The case for governance is one to promote the security, prosperity and values of spacefaring states through multilateral cooperation and collective actions to safeguard and optimise the use of space as a global commons.

Effective collection action fundamentally requires reconciliation between national and mutual, international interests. National interests are served by rules to provide reassurance that weaker players will not exploit vulnerabilities of space powers, that developing spacefaring states will behave responsibly and that rising space powers will want to join the status-quo of space as a global commons. Mutual, international interests progress when rule-based orders attract multilateral support and sustained compliance. This implies that states must provide credible reassurance that they will follow the rules, that they will not use military and technological advantages in ways that harm others and that they will support international governance arrangements and institutions, which give all states a meaningful voice in decisions that affect global security and prosperity.

By itself, the OST regime lacks formal institutional mechanisms to promote international governance for the peaceful uses of space, monitor compliance and make collective decisions about the application of rules. Albeit there are international bodies that discuss, negotiate and implement different aspects of space governance at the policy levels, like the IADC related to orbital debris and the roles of international norms and deterrence, they are all rooted in a model of self-restraint and self-regulation. As policy and law continue to evolve, approaching space governance more comprehensively in policy and more formally in law addresses challenges to the realisation of more optimal collective action, strategic stability and assurance outcomes. Such outcomes facilitate more effective space governance for the sustainable and peaceful uses of space now and in the future.

ENDNOTES

1. Eligar Sadeh (ed.), *Space Strategy in the 21st Century: Theory and Policy*, Routledge, New York/Abingdon, 2013.
2. "Special Issue: Towards a National Space Strategy", *Astropolitics*, 8 (2-3), 2010.
3. Nancy Gallagher, "International Cooperation and Space Governance Strategy", in Eligar Sadeh (ed.), No. 1.
4. The OST regime includes the body of formal international space law, principles and resolutions of international space law under the auspices of the United Nations. See United Nations Office for Outer Space Affairs, at http://www.unoosa.org (Accessed August 2016).
5. James D. Rendleman, "Strategy for Space Assurance", in Eligar Sadeh (ed.), No. 1.
6. Eligar Sadeh, "Spacepower and the Environment", in Charles D. Lutes and Peter L. Hays with Vincent A. Manzo, Lisa M. Yambrick and M. Elaine Bunn (eds.), *Toward a Theory of Spacepower: Selected Essays*, National Defense University Press, Washington DC, 2011, athttp://ndupress.ndu.edu/Portals/68/Documents/Books/spacepower.pdf (Accessed August 2016).
7. Garrett Hardin, "The Tragedy of the Commons", *Science*, 162, 1968, pp. 1243-1248.
8. See, for example, President of the United States, *National Security Strategy*, White House, Washington DC, February 2015, at https://www.whitehouse.gov/sites/default/files/docs/2015_national_security_strategy.pdf (Accessed September 2016).
9. "Special Issue: Towards a National Space Strategy", No. 2.
10. NASA, *Orbital Debris Quarterly News*, 20 (4), October 2016, at https://orbitaldebris.jsc.nasa.gov/quarterly-news/pdfs/odqnv20i4.pdf (Accessed January 2017).
11. Inter-Agency Space Debris Coordination Committee, at http://www.iadc-online.org (Accessed August 2016).
12. *Space Debris Mitigation Guidelines of the Committee on the Peaceful Uses of Outer Space*, United Nations Office for Outer Space Affairs, 2010. See United Nations General Assembly, Resolution A/Res/62/217, January 10, 2008.
13. "Report of the Scientific and Technical Subcommittee on its forty-sixth session, held in Vienna from 9 to 20 February 2009", Committee on the Peaceful Uses of Outer Space, United Nations General Assembly A/AC.105/933, March 2009.
14. See *Space Debris Mitigation Guidelines of the Committee on the Peaceful Uses of Outer Space*, No. 12.
15. *Classification of Geosynchronous Objects*, European Space Agency, February 2010; and *Classification of Geosynchronous Objects*, European Space Agency, March 28, 2015, at http://www.astronomer.ru/data/0128/ESAclassification_Issue17.pdf (Accessed July 2016).
16. "Special Issue: Towards a National Space Strategy", No. 2.
17. Nancy Gallagher, No. 3.
18. Eligar Sadeh, No. 6.
19. Eligar Sadeh, "Harmonization of Earth Observation Data", *Astropolitics* 3 (2), 005.
20. *Committee on Earth Observation Satellites Terms of Reference*, November 2013, at http://

ceos.org/document_management/Publications/Governing_Docs/CEOS_Terms-of-Reference_Nov2013.pdf (Accessed September 2016).

21. Group on Earth Observations, at http://www.earthobservations.org (Accessed July 2016).
22. Molly K. Macauley, "Earth Observations and Space Strategy", in Eligar Sadeh (ed.), No. 1.
23. *Principles Relating to Remote Sensing of the Earth from Space*, United Nations, December 3, 1986, at http://www.un.org/documents/ga/res/41/a41r065.htm (Accessed July 2016).
24. Ibid., Principle XII.
25. James D. Rendleman, No. 5.
26. P.J. Blount, "Limits on Space Weapons: Incorporating the Law of War into the *Corpus Juris Spatialis*", International Institute of Space Law Colloquium, International Astronautics Congress, Glasgow, United Kingdom, October 2008.
27. Adam E. Frey, "Defense of US Space Assets: A Legal Perspective", *Air and Space Power Journal,* December 1, 2008.
28. Roger G. Harrison, Deron R. Jackson and Collins G. Shackelford, "Space Deterrence: The Delicate Balance of Risk", *Space and Defense* 3 (1), 2009. See http://www.usafa.edu/df/dfe/dfer/centers/ecsds/defense_journal.cfm (Accessed September 2016).
29. Personal correspondence, US Air Force Space Command, October 2015.
30. Alexei Arbatov, "Preventing an Arms Race in Outer Space", in Alexandre Kalliadine and Alexei Arbatov (eds.), *Russia: Arms Control, Disarmament, and International Security*, Institute of World Economy and International Relations, 2010; and Li Bin and Nie Hongyi, "An Investigation of China-US Strategic Stability", Gregory Kulacki (trans.), Union of Concerned Scientists, May 2009. See http://www.ucsusa.org/sites/default/files/legacy/assets/documents/nwgs/Li-and-Nie-translation-final-5-22-09.pdf (Accessed January 2017. The original article was published in Chinese in *World Economics & Politics* in 2008.
31. See Karen E. House, US Public Delegate to the 63rd Session of the United Nations General Assembly, Delivered in the Debate on Outer Space (Disarmament Aspects) of the General Assembly's First Committee, October 20, 2008, *Arms Control Update*, US Delegation to the Conference on Disarmament, Geneva, US Department of State.
32. "Treaty on the Prevention of the Placement of Weapons in Outer Space, the Threat or Use of Force against Space Objects", draft, February 12, 2008.
33. Conference on Disarmament, March 9, 2015, at http://www.unog.ch/80256EDD006B9C2E/(httpNewsByYear_en)/339C2946631405E9C1257E03005F9CD4?OpenDocument (Accessed September 2016).
34. Theresa Hitchens, "Multilateralismin Space: Opportunities and Challenges for Achieving Space Security", *Space and Defense*, 4 (2), 2010. See http://www.usafa.edu/df/dfe/dfer/centers/ecsds/defense_journal.cfm (Accessed September 2016).
35. See, for example, a review of Conference on Disarmament discussions, March 9, 2015, at http://www.unog.ch/80256EDD006B9C2E/(httpNewsByYear_en)/339C2946631405E9C1257E03005F9CD4?OpenDocument (Accessed September 2016).
36. Roger G. Harrison, Deron R. Jackson and Collins G. Shackelford, No. 28.
37. Ibid.
38. "Draft Report of the Committee on the Peaceful Uses of Outer Space on the implementation of the Third United Nations Conference on the Exploration and Peaceful Uses of Outer Space (UNISPACE III)", United Nations General Assembly, A/AC.105/C.1/L.272, November 21, 2003.
39. "Committee on the Peaceful Uses of Outer Space Report on Activities Carried out in 2009 in the Framework of the United Nations Platform on Space-based Information for Disaster

and Emergency Response", United Nations General Assembly A/AC.105/955, December 23, 2009.

40. *Draft, International Code of Conduct for Outer Space Activities*, European Union, March 31, 2014, at http://www.eeas.europa.eu/non-proliferation-and-disarmament/pdf/space_code_conduct_draft_vers_31-march-2014_en.pdf (Accessed October 2016).
41. Hillary Rodham Clinton, Secretary of State, "International Code of Conduct for Outer Space Activities", Press Statement, Washington DC, January 17, 2012, at http://www.state.gov/secretary/20092013clinton/rm/2012/01/180969.htm (Accessed October 2016).

14

Beyond Outer Space Treaty – Time for New Mechanisms?

Rajeswari Pillai Rajagopalan

Introduction

The relevance of outer space in the context of social and economic development is well known, but after a gap of few decades, states are finding a greater sense of purpose in using outer space in the national security context. There has been a spurt in outer space activities in the last decade, bringing about a sharp increase in regions such as Africa, Latin America and Asia. In a domain that was once dominated by two or three great powers – the US, USSR and UK – there are today more than 60 players, including non-state actors. This has made outer space crowded, congested and contested. If the trend continues, the long-term sustainability of outer space exploration will be severely affected. These trends bring out the need for regulating outer space activities, framing new rules of engagement and ensuring safe, secure and sustainable use of outer space. Yet there are only a handful of mechanisms that guide outer space activities.

Outer Space Treaty

The Outer Space Treaty (OST) of 1967 is considered the overarching mechanism governing international space law. The Treaty was considered in the Legal Sub-Committee in 1966, and thereafter an agreement was reached in the United Nations (UN) General Assembly in the same year. The Treaty was primarily-based on the Declaration of Legal Principles Governing the Activities of States in the Exploration and Use of Outer Space, which had been adopted by the General Assembly in 1963, but included a few new provisions. The Treaty was opened for signature by the three depositary governments – the Soviet Union, UK and US – in January 1967. The Treaty entered into force in October 1967 and was guided by the following principles:

- The exploration and use of outer space shall be carried out for the benefit and in the interests of all countries and shall be the province of all mankind.
- Outer space shall be free for exploration and use by all States.
- Outer space is not subject to national appropriation by claim of sovereignty, by means of use or occupation, or by any other means.
- States shall not place nuclear weapons or other weapons of mass destruction (WMD) in orbit or on celestial bodies or station them in outer space in any other manner.
- Astronauts shall be regarded as the envoys of mankind.
- States shall be responsible for national space activities whether carried out by governmental or non-governmental entities.
- States shall be liable for damage caused by their space objects.
- States shall avoid harmful contamination of space and celestial bodies.[1]

In a nutshell, states are not permitted to place or station any WMD including nuclear weapons in outer space. The Treaty also makes it categorical that Moon and other celestial bodies shall be used only for peaceful purposes and prohibits their use for any weapon testing or setting up military bases or installations, but the Treaty does not prohibit placement of conventional weapons in orbit, thus leaving a big vacuum for states and other parties to exploit, should they decide to. Further, the Treaty prohibits any government from making a claim on a celestial bodies such as the Moon or any other planet.

Although the OST is one of the more concrete measures regulating activities in this domain, there are gaps that need to be fixed in the face of contemporary threats and challenges. For instance, many of the definitions in the OST need to be reviewed and tightened. It is believed that the definitions and their understanding have become far too expansive in their application. The need for new definitions and clarity of certain terminologies relating to space security is absolute. How does one define, for instance, terms such 'peaceful activity', or 'space weapon', or 'defensive actions in space' or even an 'astronaut'? Traditionally, an astronaut was someone who travelled to space as part of a mission. But today with Virgin Galactic pushing space tourism, will a space tourist be considered an astronaut? Another weakness is the scope and mandate of the Treaty. A treaty that came about in the 1960s is far removed from the current and future threats facing the outer space domain.[2]

Even as one battles with the deficiencies of the OST, the challenges facing outer space are growing manifold. The next section deals with the growing problems and the imperatives for an outer space regime.

The Imperative for an Outer Space Regime

Cold War rivalry between the US and USSR was a feature of the early decades of space age, and space was seen as a domain in the hands of two or three great powers. Apart from this competition, space was seen as a benign field with huge potential in the social and economic development sectors. This is nevertheless changing and states are beginning to approach space from a comprehensive national power perspective. There is increasing exploration and exploitation of space for resources, including minerals[3], meeting energy requirements (space-based solar power), operational requirement in the area of natural disasters, direction and navigational purposes and offensive military uses. Even pursuing purely civilian applications, space-based solar power, for instance, has important technological spin-offs for furthering science, technology and defence of any country.

While these are the positive shades of space uses and the domain at large, there are several risks and challenges in the domain that call for bringing about restraint and responsible behaviour. From an Indian perspective, even though India is recognised as one of the established space powers, there are more advanced military space programmes in Asia including those with anti-satellite (ASAT) capabilities, specifically China. ASAT capabilities are inherently destabilising because they can potentially be used to interfere/destroy enemy satellites, especially during times of conflict.[4] This was typically relevant in the US-Soviet context, but is now increasingly being referred to in the Asian context following China's ASAT test in January 2007. The Chinese ASAT test, after a gap of more than two decades, broke the unofficial moratorium that was in place and unleashed new anxieties in the region and beyond. The ASAT weapons' potential to disable global positioning system (GPS) satellites, thus possibly affecting GPS-based targeting and navigation systems, is considered a serious threat. The Chinese ASAT test in fact served as a wakeup call to the kind of threats and challenges that exist in India's neighbourhood.[5] While most space powers have a stated policy of civilian space programmes, there are significant gaps between rhetoric and reality. India has made significant investments in this area, and is justifiably concerned about the ASAT missiles. India's assets including the ground-based assets and associated services are worth US$ 37 bn. In fact, the Chinese ASAT tests gave way to a new debate within India on how New Delhi should be gearing up to protect its assets in outer space. The scientific, technical, military and even sections of the political community in India have talked about the need to institute appropriate measures to defend and deter its interests in outer space.

Proliferation of space technologies to state and non-state actors is another major concern. While many of the new entrants, especially in Africa and Latin America, are pursuing space for meeting their economic and developmental needs, Asia possibly has a different narrative. New entrants from Asia are entering the space domain with the security perspective dominating their discourse. This in itself has the potential to increase rivalries and conflicts.

Outer space exploration for development has become unavoidable, but it needs to be pursued in a judicious manner so as to keep outer space clean and secure. This brings into play the need to devise certain broad rules of the road for regional and international cooperation, such as in the nuclear domain. Outer space shares some common elements with the nuclear domain, but there are also important differences between the two. While space cooperation is generally seen as positive, unregulated missile and space cooperation could further regional and international insecurities. For instance, even as India has the most advanced satellite fabrication and launch capabilities in Southern Asia, external cooperation can change this. India has not forgotten the history and consequences of nuclear cooperation in its neighbourhood. While it would not like to see a repeat of such collaboration in the area of outer space in Southern Asia, given that outer space is one of the global commons with huge potential in the social and economic domains, it is difficult to restrict the flow of technology. Thus, one has to conceive measures of a different kind, and not Non-Proliferation Treaty (NPT)-like mechanisms that control the technology than its end-uses. The end-use aspect is particularly important because countries could pursue civil space technologies, which could later be diverted for military space programmes or developing long-range missile programmes. Hence, there is a need to review the existing mechanisms or frame new rules. This is particularly the case in Asia, which is already reeling under considerable insecurities, and the absence of rules and regulations can make the scenario further vulnerable.

The growing number of satellites and the types of actors is a factor as well, considering the long-term sustainability of outer space. From a mere two-three actors in the 1960s and 1970s, there has been a vast expansion in the number of space players. There are more than 60 operators, including non-state actors such as private sector players and educational institutions, among others. All of this leads to crowding and competition for the limited space. One has to recognise that space has finite resources. Unless states make a conscious effort in limiting certain type of activities, aspects of outer space will be at serious risk. In addition to the growth in the number of players, there is also growth and proliferation of small satellites – mini, micro and nano satellites, making outer space environment not only crowded and congested but also rendering the process of detection and tracking extremely challenging.

Growing interface between cyberspace and outer space is also emerging as a critical challenge to outer space sustainability. One of the most common types of cyberattacks on space systems is jamming, although many do not consider it as a cyberattack. Jamming occurs when a stronger signal is able to override and supress the signal from a satellite.[6] This is not in the realm of science fiction – for instance, between 2010 and 2012, North Korea had reportedly jammed South Korean GPS signals for several days, impacting upon its planes, ships and even personal devices.[7]

There also increasing debates between weapons in outer space and ground-based systems that can be used for targeting assets in outer space. One can be reasonably certain that no country in its rightful mind is going to place weapons in outer space, but the possibility of states using kinetic killer anti-satellites weaponry to destroy or damage assets in outer space cannot be ruled out. A few countries have suggested that placement of weapons in outer space is the greatest challenge today and that the dangerous consequences of ASAT missiles are underplayed. For countries like India, the biggest challenge is not WMD or conventional weapons in outer space, and the focus is on ASAT-like capabilities.

Even as spectrum allocation is becoming a serious challenge, there is a more important and immediate threat of space debris which if unattended could lead to a scenario wherein use of outer space even for the current generation would become problematic.

Increasing population of space debris is already risking the uninterrupted use of outer space. Ecuador lost its one and only satellite, a CubeSat named NEE-01 Pegasus, to a Russian piece of space debris. It is estimated that there

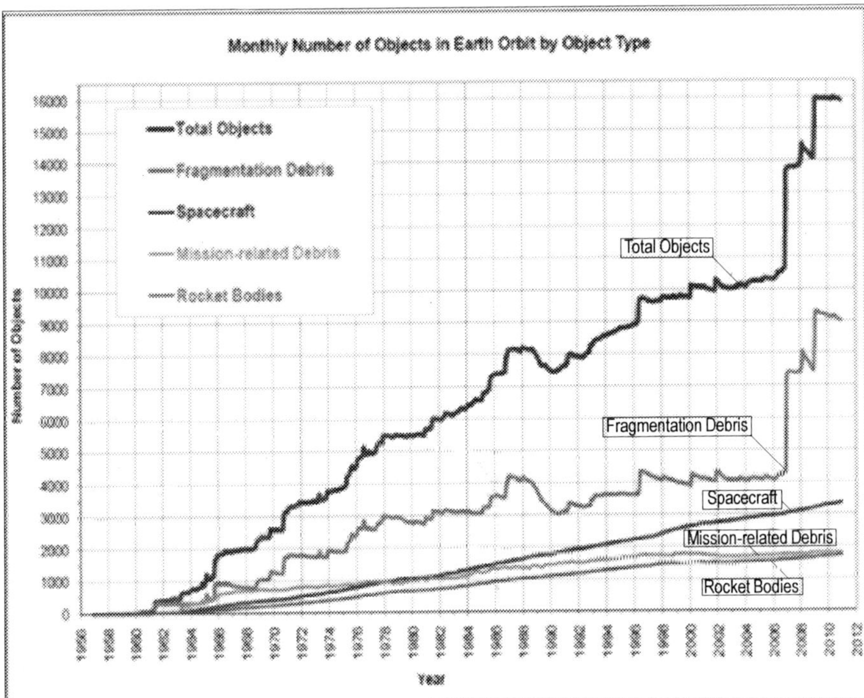

Monthly Number of Cataloged Objects in Earth Orbit by Object Type. This chart displays a summary of all objects in Earth orbit officially cataloged by the U.S. Space Surveillance Network. "Fragmentation debris" includes satellite breakup debris and anomalous event debris, while "mission-related debris" includes all objects dispensed, separated, or released as part of the planned mission.

Figure 1. Monthly Number of Objects in Earth Orbit by Object Type

are more than 21,000 pieces of space debris that are larger than 10 cm, approximately 500,000 pieces of size between one and 10 cm and more than 100 million reported to be smaller than 1 cm. Tracking and detection of the smaller pieces of debris are a big problem. India's concerns in this regard are clear too. India's civil space organisation, Indian Space Research Organisation (ISRO) has had to delay its satellite launches several times in recent years due to potential satellite collision incidents.

Another important challenge in the area of outer space security and sustainability come from the potential interference in radio frequency and spectrum allocation. Crowded spectrum could itself lead to radio frequency interference even if unintentional in nature. In addition, there are also phenomena such as space weather and other natural events that could create frequency interference. But there are also states that are increasingly developing capabilities for jamming and frequency interference in order to inflict damage on an adversary's ability to rely on satellite capabilities.

This leads one to the next problem of space traffic management. Just as there are rules regulating traffic on earth, there is a need to manage and regulate traffic in outer space. The number of small satellites and satellite operators around the world have gone up rapidly, making a crowded environment even more congested. For instance, the Low Earth Orbit is crowded with active satellites, millions of pieces of debris from dead satellites, spent rocket boosters and other stray hardware. This increases the potential for collision between various space objects.

While each of these issues are significant on its own, the state of the regime is the biggest challenge facing the outer space today. Disagreements among major space players to tackle the challenges have essentially contributed to the slow development of an outer space regime. Several different factors have contributed to this current scenario. For one, proliferation of space technology to a larger group of actors makes the process of finding agreement extremely challenging. Earlier, when technology was in the hands of two or three major powers, it was easier to reach a consensus because these countries also had an inherent interest in controlling the spread of technology. This is more complicated now. Another factor is the changing global power equations. The shifting balance of power with the relative decline of the US, or at least the perception of it, has contributed a great deal to the US inability to push through an effective space regime. On any important global security issue, there is the US-led West and the Russia-China duo that have emerged, countering each other's position. Unfortunately, it is unlikely that this phase of disagreement is going to go away in the near future. Nevertheless, given the growing number of imminent challenges, states have to make a conscious effort in devising measures that are minimally acceptable to each of the established space players.

Other Existing Mechanisms

In addition to the Outer Space Treaty, there are a few existing mechanisms that merit attention. These include the UN Committee on the Peaceful Uses of Outer Space (COPUOS), Prevention of an Arms Race in Outer Space (PAROS), Treaty on Prevention of the Placement of Weapons in Outer Space and of the Threat or Use of Force against Outer Space Objects (PPWT), International Code of Conduct for Outer Space Activities (ICoC) and UN Group of Governmental Experts (GGE). Each of these has its own strengths and weaknesses, which are detailed below.[8]

COPUOS has been one of the best UN platforms to manage issues around the peaceful uses of outer space, as the name suggests. Established in 1959 through a UN General Assembly resolution in order to streamline international cooperation and further research and dissemination of information relating to outer space, it functions primarily through two subcommittees – the Scientific and Technical Subcommittee and the Legal Subcommittee. The decisions of the COPUOS are enforced by the UN Office of Outer Space Affairs (OOSA). But the Committee suffers from one major weakness that an entire array of activities that fall in the area of military and security are not part of its mandate. The fact that the COPUOS has a limited mandate has been exploited by several states in sending multiple and contradictory messages about their intentions and programmes. However, in June 2007, the COPUOS adopted Debris Mitigation Guidelines, which are not strictly within the peaceful uses mandate of the COPUOS. So to some extent, the criticism about the limited mandate of the COPUOS is addressed here. Many countries prefer the COPUOS over other such platforms because a larger number of states are party to it and therefore it is seen as more inclusive.

Just as the COPUOS is meant to debate on all peaceful uses of outer space, the Conference on Disarmament (CD) in Geneva is the principal forum where non-proliferation and security-related aspects of outer space have been debated. But the CD has stagnated for two decades now to the point that there is not even an agreement on the agenda. Given the stalemate that the CD is characterised with, state parties have been trying to find innovative ways of using other existing platforms to make progress on space security issues. Therefore, if the COPUOS could debate debris mitigation issues and come out with a document that has the support of all nations, may be it will replace the CD as the venue for pertinent space security issues as well.

Similarly, PAROS negotiations have not gone very far despite a resolution passed by the UN General Assembly in 1981; and three decades later, there has been no progress. This belies the fact that there is a near global consensus that outer space should be used only for peaceful and civilian purposes. It is essentially the political disagreements among major powers that have come in the way of making progress on PAROS. There is also the criticism that PAROS is excessively focused on arms race and neglects some other more critical issues such as space debris and the trend towards weaponisation of outer space.[9]

Meanwhile, Russia and China have tried to pursue new treaty-like mechanisms to limit the trend towards weaponisation of outer space. In 2008, the governments of Russia and China submitted a draft PPWT to the CD in Geneva. The Treaty could not gather much support for several reasons including the fact that it was non-verifiable. Also the excessive focus on weapons in outer space was noted as problematic. Further, the Treaty does not mention space debris or ASAT capabilities or soft kill weapons such as lasers to damage or destroy satellites. In the backdrop of the debates on the code of conduct, Russia and China submitted a new version of their draft Treaty in June 2014, but many of the weaknesses from the earlier draft are seen in the current version as well.[10] There are also several definitional issues that make the Treaty less appealing. For instance, the Treaty attempts to define outer space objects in a superfluous manner while the Liability Convention and the Rescue Agreement have already defined the same. This is further problematic because while framing new rules and treaties, effort should be made to integrate those new measures into the existing international space law rather than come up with a new understanding.

After being faced with significant challenges in the area of rule-making particularly those of legally binding mechanisms, efforts are being made today to start with broad international political agreements which are relatively easier for countries to conclude. These include transparency and confidence building measures (TCBMs) such as codes of conduct and the GGE under the UN. Codes of conduct are a good initiative because they codify certain principles, rules, best practices while contributing to TCBMs and complimenting existing arrangements on outer space activities. One of the recent measures that gained reasonable traction was the European Union (EU)-proposed ICoC. The ICoC is quite comprehensive in terms of the issues covered and the suggested norms of acceptable behaviour to regulate outer space activities, but it suffered significant setback on account of process issue.[11] It was alleged that the EU did not engage in prior consultation with even the established space powers, thus failing to get the buy-in of these countries.

In the face of increasing disagreements among major powers, the GGE is another political initiative that has been contemplated in recent times. The GGE has a reasonable level of buy-in because it is established within the UN. Nevertheless, these are smaller mechanisms with limited membership, and therefore the question of how inclusive the GGE is remains an issue. There are also other issues with the GGE because its recommendations are not binding. However, some governments have been proactive in building on the recommendations and submitting them to the General Assembly as resolutions. This, in a sense, rectifies the problem of the lack of an inclusive platform, to some extent.

The increasing disagreements among major powers is also a reflection of the shifting balance of power and Asian uncertainties. Proliferation of space technology and shifting balance of power equations have impacted upon the exercise of international rule making.

Need to Go Beyond OST – Time for New Mechanisms?

Under the above-mentioned circumstances of growing threats and lack of consensus, major spacefaring nations have to find ways of figuring out what those minimally acceptable standards may be in order to define certain norms of responsible behaviour. Reviewing and fine-tuning the OST is the ideal situation, but the current political scenario does not appear ready to engage in any meaningful endeavour in this regard. Therefore, it is important for states to consider a few practical means to allay some of the prevailing sense of rivalry and competition in order to develop predictable and sustainable options.

Given that the political disagreements have become the biggest impediments in developing the necessary consensus among states, one of the important steps might be to strengthen dialogue with the objective of encouraging openness, greater transparency and information sharing. The need to institute multiple levels of dialogue and interaction for broader understanding of the different orientations and objectives of space programmes and policies is real. This could usher in the much needed confidence and reassurance in engaging with each other for the larger good of the outer space domain.

Also, India and other established spacefaring powers need to contemplate on the kind of measures that need to be instituted. For one, given the dual-use nature of space technology, NPT-like mechanisms are unlikely to work because control of the spread of technology is not feasible. On the other hand, space technology needs to be dealt with mechanisms such as the Chemical Weapons Convention that restrict the end use than control of technology.

The debate between soft law and hard law has also continued for close to a decade now. Can TCBMs such as the Code of Conduct and GGE emerge as possible solutions in the near term?[12] The TCBMs are essentially voluntary, non-legal measures that promote mutual trust, encourage cooperation and transparency while reducing wariness, competition and misperceptions as a means to prevent intended or unintended conflicts. The TCBMs need to be pursued with greater vigour because they are an intermediate measure between a functional need and a binding instrument. They essentially serve the purpose of a bridge between the idea of an effective instrument and the fructification of one. The TCBMs in the form a code and GGE could start by identifying a few pragmatic, near term steps to improve transparency and openness. They will also provide forums and platforms to talk to each other, build up the confidence in each other and help deal with some of the political problems that come in the way of building a regime.

But TCBMs also suffer from some disadvantages. Since they are not legal measures, what if states break their political commitment? And how does one enforce these measures? Because they are non-verifiable, they could be broken with no penalties although it is easier to reach agreements on such measures domestically. Political commitments cannot guarantee responsible behaviour, but the TCBMs could create international peer pressure on states to follow through and not walk back on commitments. Moreover, not being part of them or violating commitments after signing on to them may make states subject to normative pressures. In a competitive environment such as in Asia, China cheating on a commitment would mean that India could do the same.[13]

Nevertheless, the TCBMs are gaining support for a number of reasons such as growing challenges, absence of an effective outer space regime and the need for more confidence building measures (CBMs). Therefore, the need of the hour is to start with the least controversial and minimal standards of responsible behaviour. These are important in the context of tackling several challenges such as space debris, orbital collision, promoting space situational awareness (SSA) and planetary defence. The ability to keep track of space objects, debris, space weather, including predicting orbital collisions, detecting launches of new space objects, predicting re-entry of space objects into the atmosphere and detecting threats and attacks on spacecraft have become critical for maintaining uninterrupted access to outer space. The importance of planetary defence cannot be emphasised enough – a meteor hit in Russia and Asteroid 2012 DA 14's close miss in 2013 demonstrate the reality of threat from asteroids and meteors.[14] Thus, states have to make all-out efforts to pursue both binding and non-binding instruments.

Multiple measures have been so far debated without much success but states have to be reminded that space debris, for instance, does not make a distinction between different states' space assets. Therefore, the responsibility to keep the outer space clean, safe and secure has to be taken up by every state, especially the established spacefaring powers. Legally binding mechanisms are ideal and desirable. However, the prospect of consensus among major powers in identifying challenges and introducing solutions seems distant.

What do these mean for India and how should New Delhi respond? One, the political leadership has to take a greater ownership of the domain and define space from a commercial and national security perspectives. Two, with the global debate on framing new rules of the road, India needs to come up with a considered and comprehensive approach – a similar approach is seen in the area of nuclear and WMD security. Prime Minister Narendra Modi's recent proactive approach at the 2016 Nuclear Security Summit is a case in point. India could possibly work with other established spacefaring powers in taking the lead in this regard.

ENDNOTES

1. United Nations Office for Outer Space Affairs, "Treaty on Principles Governing the Activities of States in the Exploration and Use of Outer Space, Including the Moon and Other Celestial Bodies: General Assembly Resolution 2222 (XXI), Annex", Resolution adopted on December 19, 1966 and Treaty signed on January 27, 1967), at http://www.unoosa.org/oosa/en/ourwork/spacelaw/treaties/outerspacetreaty.html.
2. Rajeswari Pillai Rajagopalan, "The International Code of Conduct and Space Sustainability", in C. Al-Ekabi, B. Baranes, P. Hulsroj, A. Lahcen, (eds.), *Yearbook on Space Policy 2014: The Governance of Space*, Springer, Vienna, 2015, pp. 229-41.
3. Chaitanya Giri, "Mining in Space – The Next Frontier?", *The Diplomat*, July 16, 2013, at http://thediplomat.com/2013/07/mining-in-space-the-next-frontier/?allpages=yes.
4. Rajeswari Pillai Rajagopalan, "Debate on a Space Code of Conduct: An Indian Perspective", Presentation at the United Nations Institute for Disarmament Research (UNIDIR) Annual Space Security Conference, "Space Security Conference 2012: Laying the Groundwork for Progress", March 29-30, 2012, at http://www.unidir.ch/files/conferences/pdfs/an-indian-perspective-en-1-688.pdf
5. Rajeswari Pillai Rajagopalan and Arvind John, "A New Frontier: Boosting India's Military Presence in Outer Space", Occasional Paper No. 50, Observer Research Foundation, January 2014, at http://dhqxnzzajv69c.cloudfront.net/wp-content/uploads/2014/02/Occational-Paper_50.pdf.
6. "The Space Industry Wakes up to Cyber Threat", Interview with Denis Bensoussan, Head of Space at Beazley, July 07, 2014, at http://www.lloyds.com/news-and-insight/news-and-features/market-news/industry-news-2014/the-spaceindustry-wakes-up-to-the-cyber-threat. Also see Maj. Brian Garino, USAF, and Maj. Jane Gibson, USAF, "Space System Threats", AU-18: Space Primer, US Air Command and Staff College, US Air University, 2009, Chapter 21, at http://www.au.af.mil/au/awc/space/au-18-2009/.
7. Martyn Williams, "DPRK Jamming GPS Signals, Says Seoul", *North Korea Tech*, May 3, 2012, at http://www.northkoreatech.org/2012/05/03/dprkjamming-gps-signals-says-seoul;

"North Korea Jamming' Hits South Korea Flights", *BBC* News, May 2, 2012, at http://www.bbc.com/news/world-asia-17922021; Shaun Waterman, "North Korean Jamming of GPS Shows System's Weakness", *The Washington Times,* August 23, 2012, at http://www.washingtontimes.com/news/2012/aug/23/ northkorean-jamming-gps-shows-systems-weakness/?page=all.

8. On the strengths and weaknesses of many of these measures, see Paul Meyer, "Putting Cooperation back into Space Security", Presentation at the 10th UN Workshop on Space Law, Contribution of Space Law and Policy to Space Governance and Space Security in the 21st Century, September 5, 2016, Vienna, Austria, at http://www.unoosa.org/pdf/SLW2016/Opening/3._Meyer_-_SpaceCoopUNOOSASep52016.pdf.
9. Paul Meyer, "The Conference on Disarmament and the Prevention of an Arms Race in Outer Space", *The CD Discussion Series,* April 2011, at http://www.unidir.org/files/publications/pdfs/the-conference-on-disarmament-and-the-prevention-of-an-arms-race-in-outer-space-370.pdf.
10. Michael Listner and Rajeswari Pillai Rajagopalan, "The 2014 PPWT: A New Draft but with the Same and Different Problems", *Space Review,* August 11, 2014, at http://www.thespacereview.com/article/2575/1.
11. Ajey Lele (ed.), *Decoding the International Code of Conduct for Outer Space Activities,* Pentagon Press, New Delhi, 2012, at http://www.idsa.in/system/files/book/book_OuterSpaceActivities.pdf; Rajeswari Pillai Rajagopalan, "The Space Code of Conduct Debate: A View from Delhi", *Strategic Studies Quarterly,* Spring 2012, at http://www.au.af.mil/au/ssq/2012/spring/rajagopalan.pdf; Rajeswari Pillai Rajagopalan, "Debate on Space Code of Conduct: An Indian Perspective", Occasional Paper No. 26, Observer Research Foundation, October 2011, at http://www.orfonline.org/wp-content/uploads/2011/10/OccasionalPaper_26.pdf.
12. Guoyu Wang, "TCBMs and the Outer Space Treaty", Presentation at the third ASEAN Regional Forum (ARF) Workshop on Space Security, Beijing, China, November 30, 2015, at http://www.unidir.ch/files/conferences/pdfs/tcbms-and-the-outer-space-treaty-en-1-1100.pdf.
13. Rajeswari Pillai Rajagopalan, "Relationship of Norms of Behaviour with TCBMs and Treaties, in Particular the Legal and Political Advantages and Disadvantages of the Different Types of Instruments", Presentation at the UNIDIR Conference, "Facilitating the Process for the Development of an International Code of Conduct for Outer Space Activities", Malaysia, December 12, 2012, at http://www.unidir.org/files/medias/pdfs/rajeswari-pillai-rajagopalan-eng-0-502.pdf.
14. Rajeswari Pillai Rajagopalan, "Asteroids, Meteors and Planetary Defence", *Commentary,* Observer Research Foundation, February 18, 2013, at http://www.orfonline.org/research/asteroids-meteors-and-planetary-defence/.

15

The Future of the Outer Space Treaty

Ram S. Jakhu

Introduction

The Outer Space Treaty,[1] the hallmark of global space governance, is a product of the post-Second World War period and essentially of Cold War geo-politics. So far, it has undoubtedly succeeded in maintaining international space order in accordance with the wishes of its authors and the central objectives of this Treaty. However, nobody can say anything about the fate of the Treaty with any certainty. In other words, it is extremely difficult, if not impossible, to predict the precise future of the Treaty, especially because of the highly volatile current geo-political situation in the world, the increasing number of space activities and space actors, the rapidly changing nature and complexity of space operations and numerous other factors impacting on global governance, and global space governance in particular. In this short chapter, I intend to briefly describe some possible scenarios about the future of the Treaty and a few key implications thereof.

Four Possible Scenarios

1. *No Change*

It appears more likely that there will be no change in the status quo of the Outer Space Treaty, at least in the near future. There are several reasons for this. Firstly, there currently is no serious imminent threat to the efficacy or prospect of the severe violation of the key principles of the Treaty. Secondly, the Treaty does not contain any procedure for its formal review. Therefore, if they wish to undertake such a step, the states parties would need to initiate the convening of an assembly of parties, or seek a resolution of the United Nations Committee on Peaceful Uses of Outer Space (UNCOPUOS), and in turn eventually the UN General Assembly (UNGA). The requirement of consensus for the adoption of a decision at the UNCOPUOS would make the approval of a resolution to amend the Outer Space Treaty very challenging, as some states are not in favour of any change in the Treaty, primarily because this may open a Pandora's Box whereby there may be unpredictability with regard to ensuing developments.

Consequently, the Treaty would continue operating as it is, even though its provisions are increasingly becoming insufficient and inadequate to effectively regulate the fast emerging issues. Sadly, some states may simply start ignoring or interpreting differently the provisions which they consider to be not in their national interest. We are now already witnessing this as, in particular, the major space powers are seeking to develop increasingly dangerous and potentially destructive space-related weapons, which is in every sense contrary to the underlying principles of the Treaty.

If this trend becomes even more common, one may see the creation of state practice that could 'change' the original intent or meaning of some provisions of the Treaty. Article 31(3) of the 1969 Vienna Convention on the Law of Treaties specifies that, in the interpretation of provisions of a treaty, "any subsequent practice in the application of the treaty which establishes the agreement of the parties regarding its interpretation, may be taken into consideration".[2] One may note the occurrence of such a change in the meaning of term "peaceful use" under the Outer Space Treaty. Normally, the term "peaceful" means non-military; however, within the context of the Outer Space Treaty, it has become accepted that "peaceful" includes "military", while still excluding "aggressive" uses, as almost all states have started following the practice initiated by the US and former Soviet Union in using outer space for a variety of uses related to the military.

2. *Settlement of Disputes Arising under the Outer Space Treaty*

Another way the Outer Space Treaty may be affected is through some form of adjudication or dispute resolution based on the Treaty. Some states parties may attempt to seek, through international judicial tribunals, the settlement of disputes arising under the provisions of the Outer Space Treaty, if they believe that other state(s) are not in compliance with its specific provisions (i.e. committing an internationally wrongful act or omission). However, the issue of jurisdictional competence of the concerned judicial tribunals, including the International Court of Justice (ICJ),[3] would become a serious barrier, since the Outer Space Treaty does not contain any provision related to the settlement of disputes through peaceful means.

General international law obliges states to settle their disputes by peaceful means, but leaves it to the states to determine specific mechanisms for fulfilment of this legal obligation.[4] A state (alone or in collaboration with other states) that is not a party to the Outer Space Treaty may take steps towards judicial determination pursuant to the provisions of Article 48 of the 2001 Draft Articles on Responsibility of States for Internationally Wrongful Acts.[5]

Nevertheless, the aggrieved State(s) may successfully secure a UNGA resolution seeking an advisory opinion by the ICJ on specific provisions of the Treaty and/or actions of some states parties to the Treaty.[6] The primary candidate for seeking such an advisory opinion could be Article II of the Outer Space Treaty, which prohibits appropriation of outer space and celestial bodies, particularly in view of the 2015 US SPACE Act that purports to allow private property rights over space natural resources to American companies.[7]

Another possible issue for an advisory opinion could be the clarification of provisions of Article IX of the Outer Space Treaty in circumstances where a state party intentionally creates an extensive amount of space debris or widely contaminates outer space or any celestial body.

3. *Modification and/or Expansion of Some Provisions of the Outer Space Treaty*

Since its adoption, the Outer Space Treaty has not been amended,[8] although it has been supplemented by four other international agreements[9] that were negotiated and adopted through the UNCOPUOS. These agreements are, in essence, based on elaborations of some provisions of the Outer Space Treaty. However, since 1979, the negotiation of such internationally binding instruments has almost stopped. There seems to be a very little, if any, political will, particularly on the part of major space powers, to enter into additional international agreements. Instead, there is a strong trend towards the adoption of non-binding (soft-law) resolutions or guidelines that might be considered to supplement, or clarify the meaning of some provisions of the Outer Space Treaty.[10] The provisions of these instruments might ultimately become norms of customary international law, thereby either supplementing or modifying the provisions of the Outer Space Treaty. However, declarations attached to some of these instruments expressly proclaiming their non-binding nature, and the international doctrine of the "persistent objector", would prevent soft-law documents from becoming rules of customary international law.[11]

4. *"Life without the Outer Space Treaty": Withdrawal by Major Space Power(s)*

It may seem unlikely, but it is conceivable that a major space power may decide to withdraw from the Outer Space Treaty if it considers that the Treaty is no longer in the state's national interest. Such a scenario may become more probable as the current trend towards nationalistic populism expands. One commentator rightly pointed out that "Brexit and Trump prove that the world is not globalising. It's localising".[12] Unfortunately, it is expected that other states, possibly France, Germany and Italy in Europe, will follow the lead in this process taken by the UK and US.[13]

One of the first victims of this trend will be the key international treaties and institutions which have been the hallmark of progressive internationalism after the Second World War. In a climate of foreign policies that are potentially isolationistic and nationalistic, the Outer Space Treaty may become one such international agreement leading spacefaring states may opt to leave. This may especially be the case to the extent that provisions of the Treaty are seen as serious roadblocks in pursuing economic interests, such as the exclusive national exploitation of space natural resources by some states and their private companies, and undermining national security interests, such as in the implicit prohibition on the deployment of weapons for use in and from space.

Undoubtedly, the Outer Space Treaty is the most important foundation of the legal order for outer space. Even so, it is simply an international treaty which states parties are entitled to (and could) repudiate by giving a notice of withdrawal that would become effective one year thereafter.[14] Both the logical opinion of credible space law experts and common sense of informed people will require that such an important treaty must be strengthened, or at least not demolished. However, one wonders if, in light of Brexit and Trump's ascension to the highest position in the most powerful country in the world,[15] common sense will continue to prevail and whether a balanced and fair media, globally-known intellectuals, and smart and experienced political leaders still have the influence to safeguard the Outer Space Treaty, which is, as noted above, a vital cornerstone of the global space governance.

Even if one does not believe in Murphy's law; i.e. "*whatever can happen will happen*", in the context of current and emerging geopolitical trends it would be prudent to postulate the possible demise (or significant weakening) of the Outer Space Treaty. Is it possible to conceive of "life without the Outer Space Treaty"? Would there be a complete breakdown of law and order governing activities in outer space and, if so, what possible mechanisms could be utilised to dampen the devastating impact of such a dismantling of the governance system that has survived and persisted in the face of geo-political changes and upheavals in the past five decades?

Admittedly, withdrawal from the Outer Space Treaty by one or two major space powers will not be the end of the world, but it is conceivable that it could trigger the Doomsday Clock to advance from three to two minutes to midnight.[16] Moreover, it would create chaotic situation in the current global space governance system and result in serious turmoil in the global political and strategic sphere.

In this brief chapter, it is not possibly to discuss all the significant implications of withdrawal from the Outer Space Treaty on global space governance. However, the following important points should be noted in this regard:

> Subject to any other relevant principles of international law, a state which is no more a party to the Outer Space Treaty is not bound by its provisions, and thus, among other things, may believe that it is (legally) permitted to (a) place weapons of mass destruction in outer space;[17] (b) appropriate (or allow its private entities to appropriate) by declaration or use a part of outer space or a celestial body or their natural resources;[18] (c) allow its private companies unfettered freedom to carry out space activities without any supervision, safety standards and due regard to the interests of other states;[19] (d) establish military basis or test weapons on celestial bodies;[20] and (e) not pay compensation to the victims of damage caused by space objects belonging to its private entities.[21]

It is not difficult to imagine the legal and political consequences of such actions on the part of the withdrawing state. It is a well-accepted fact that several key principles of the Treaty have become norms of customary international law (and some are even *jus cogens*).[22] If a major spacefaring state withdraws from the Outer Space Treaty, though these basic principles will apply – and will continue applying – to all states whether or not parties to the Outer Space Treaty, they may be perceived as becoming weak and limited in application.

One can expect that the rationale for withdrawing from the Outer Space Treaty signals that the withdrawing state would not fully comply with certain norms of customary international law applicable to outer space – or, indeed, they may not believe that they represent binding customary international law at all. Further, a withdrawal on the part of a major spacefaring state will render the international space legal regime unstable and unpredictable. Yes, one may argue that an aggrieved state (or a group of states) may invoke the principles of state responsibility for actions contrary to those provisions of the Outer Space Treaty that have become part of customary international law. However, as noted above, it will be a serious challenge to bring a case before an international judicial tribunal for the determination of state responsibility.

The more serious problem, it is believed, will be international political turmoil and uncertainty. When multilateral diplomacy or the international rule of law fails, unilateral national actions emerge. Conversely, in order to take unilateral actions, or to develop national military capabilities, or to secure access to resources, it is not difficult to foresee the suppression or discontinuation of multilateral constraints, rules and diplomacy.

One may remember the global political reaction to the 2001 unilateral withdrawal by the US from the 1972 bilateral Anti-Ballistic Missile (ABM) Treaty,[23] which was considered as a "cornerstone of strategic stability".[24] After the termination of the ABM Treaty, both the US and Russian Federation started the increased deployment of their weapons, which invigorated an arms race.

With the benefit of hindsight, throughout history the two major reasons – premises – for war or conflicts are fear of, or desire to exercise domination over, others and motivations of national interest to gain access to resources. Both these grounds can be the basis for withdrawing from the Outer Space Treaty as well as national action following such withdrawal. The provisions of the Outer Space Treaty, particularly in limiting the deployment of weapons of mass destruction and imposing a comprehensive ban on the appropriation of outer space and celestial bodies, have been carefully designed to favour multilateralism and underplay nationalistic and self-interests motivations in the exploration and use of the global commons that is outer space. Being not bound by the provisions of the Treaty, a state may be expected to pursue national actions for the exclusive exploitation of space natural resources and to develop and deploy weapons in outer space; consequently, there would emerge serious international conflicts and a space arms race. The 2016 session of the UNCOPUOS witnessed a serious international negative reaction to the 2015 US SPACE Act that seeks to grant national private property rights over space natural resources.[25]

In order to avoid this scenario becoming a reality, other major spacefaring nations and space middle powers would have to make concerted diplomatic efforts to keep the integrity and effectiveness of the Outer Space Treaty intact. After all, they would the ones whose vital interest could suffer the most.

Conclusion

The scenarios about the future of the Outer Space Treaty, as briefly described above, are not definitive. However, the probabilities of their occurrence, in various forms and at different times, cannot be overlooked. The efficacy and continuation of the Outer Space Treaty, even in its current form, must not be taken for granted. This foundational piece of international legislation is indispensable for the reaping and equitable sharing of unimaginable benefits derived from outer space for humanity on Earth. The third scenario, which favours the strengthening and expansion of the existing global space governance system to accommodate changing and emerging needs and interests, provides the best course of future action and development for the international community, especially as far as the major space-faring nations and space middle powers are concerned.

However, the possibility of the occurrence of the fourth scenario, though the least desirable situation, must not be discounted. "Life without the Outer Space Treaty" would have very serious implications for all nations and peoples–consequences and implications that must be very carefully postulated and understood, so that humanity may be prepared for such an eventuality. It is hoped that, true to the intentions of the original drafters of the Treaty, and true to the objects and purposes behind this significant legislation that has maintained international peace and stability and ensured the orderly development of space activities in the past five decades, in the interest of all humankind and for the sake of the multilateral world order, the Outer Space Treaty will continue to play a significant role in guiding present and future space activities.

Acknowledgement: The author acknowledges with sincere gratitude the help provided by Mr. Kuan-Wei (David) Chen and Prof. Steven Freeland in reviewing and providing valuable comments and edits on the earlier draft of this chapter that improved its quality. As always and notwithstanding this invaluable help, the author remains exclusively responsible for any errors contained in this paper: Ram S. Jakhu.

ENDNOTES

1. *Treaty on Principles Governing the Activities of States in the Exploration and Use of Outer Space, Including the Moon and Other Celestial Bodies*, January 27, 1967, 610 UNTS 205, 18 UST 2410, TIAS No 6347, 6 ILM 386 (entered into force on October 10, 1967) [*Outer Space Treaty*]. There are 104 ratifications and 25 signatures (as of December 31, 2016).
2. *Vienna Convention on the Law of Treaties*, May 23, 1969, UN Doc A/Conf.39/27, 1155 UNTS 331, 8 ILM 679 (1969), 63 AJIL 875 (1969) (entered into force on January 27, 1980).
3. See generally *Statute of the International Court of Justice*, June 26, 1945, 3 Bevans 1179, 59 Stat 1031, TS 993, 39 AJIL Supplement 215 (entered into force October 24, 1945), and particularly Art. 36.
4. UN Charter, Art. 33; Principle 2 (2) of *Declaration on Principles of International Law Concerning Friendly Relations and Co-operation among States in Accordance with the Charter of the United Nations*, GA Res. 2625 (XXV), UNGAOR, 25th Session, Supplement No. 28, UN Doc A/8028 of October 24, 1970; *Aerial Incident of 10 August 1999 (Pakistan v. India)*, [2000] ICJ Rep 12 (para. 53); *UN Handbook on the Peaceful Settlement of Disputes between States*, UN Doc OLA/COD/2394, United Nations, New York, 1992, at www.un.org/law/books/HandbookOnPSD.pdf.
5. *Draft Articles on Responsibility of States for Internationally Wrongful Acts*, International Law commission (ILC), 53 UN GAOR Supplement (No. 10) at 43, UN Doc A/56/83 (2001), Art. 48:
   1. Any State other than an injured State is entitled to invoke the responsibility of another State in accordance with paragraph 2 if:
      (*a*) the obligation breached is owed to a group of States including that State, and is established for the protection of a collective interest of the group; or
      (*b*) the obligation breached is owed to the international community as a whole.
   2. Any State entitled to invoke responsibility under paragraph 1 may claim from the responsible State:
      (*a*) cessation of the internationally wrongful act, and assurances and guarantees of non-repetition in accordance with article 30; and
      (*b*) performance of the obligation of reparation in accordance with the preceding articles, in the interest of the injured State or of the beneficiaries of the obligation breached.
6. *Statute of the International Court of Justice*, No. 3, Art. 65.
7. *US Commercial Space Launch Competitiveness Act*, Pub L 114-90, Title I, *Spurring Private Aerospace Competitiveness and Entrepreneurship Act of 2015* [*SPACE Act*], particularly s 51303.
8. *Outer Space Treaty*, Art. XV, allows any state party to propose amendments to this Treaty.
9. They are: the *Agreement on the Rescue of Astronauts, the Return of Astronauts and the Return of Objects Launched into Outer Space* (Rescue Agreement), adopted by the General Assembly in its resolution 2345 (XXII), opened for signature on April 22, 1968, entered into force on December 3, 1968 – there are 95 ratifications, 24 signatures, and two acceptance of rights and obligations (as of December 31, 2016); the *Convention on International Liability for Damage Caused by Space Objects* (Liability Convention), adopted by the General Assembly in its resolution 2777 (XXVI), opened for signature on March 29, 1972, entered into force on September 1, 1972 – there are 89 ratifications, 22 signatures, and three acceptances of rights and obligations (as of December 31, 2016); the *Convention on Registration of Objects Launched into Outer Space* (Registration Convention), adopted by the General Assembly in its resolution 3235 (XXIX), opened for signature on January 14, 1975, entered into force

on September 15, 1976 – there are 63 ratifications, four signatures, and two acceptances of rights and obligations (as of December 31, 2016); the *Agreement Governing the Activities of States on the Moon and Other Celestial Bodies* (Moon Agreement), adopted by the General Assembly in its resolution 34/68, opened for signature on December 18, 1979, entered into force on July 11, 1984 – there are 17 ratifications and four signatures (as of December 31, 2016).

10. Some of these instruments are: the UN *Principles Relating to Remote Sensing of the Earth from Outer Space*, adopted by the UN General Assembly under Resolution 41/65 on December 3, 1986; the UN *Principles Relevant to the Use of Nuclear Power Sources in Outer Space,* adopted by the UN General Assembly under Resolution 47/68 on December 14, 1992; the *Declaration on International Cooperation in the Exploration and Use of Outer Space for the Benefit and in the Interest of All States, Taking into Particular Account the Needs of Developing Countries*, adopted by the UN General Assembly under Resolution A/RES/51/122 on December 13, 1996; the *Application of the Concept of the "Launching State"*, adopted by the UN General Assembly under Resolution A/RES/59/115 on January 25, 2005; the COPUOS *Space Debris Mitigation Guidelines*, endorsed by UN General Assembly under Resolution A/RES/62/217 (para. 26) of January 10, 2008; the Hague Code of Conduct against Ballistic Missile Proliferation, 2002; Resolution of the UN General Assembly on *Prevention of an Arms Race in Outer Space*, UN Doc. A/RES/71/31 of December 09, 2016; Resolution of the UN General Assembly on *No first placement of weapons in outer space*, UN Doc. A/RES/71/32 of December 09, 2016; Resolution of the UN General Assembly on *Transparency and confidence-building measures in outer space activities*, UN Doc A/71/42 of December 9, 2016.
11. *Anglo Norwegian Fisheries Case* (UK vs Norway), [1951] ICJ Rep. 3.
12. Chris Skinner, "Brexit and Trump Prove that the World Is Not Globalizing", IRIS, November 11,2016, at http://www.iris.xyz/fintech/brexit-and-trump-prove-world-not-globalizing.
13. Brian O'Connor, "President Trump and Brexit Are Only the Start", *Time*, November 10, 2016, at http://time.com/4566638/trump-brexit-france-germany/; Julian Coman, "First Brexit Then Trump. Is Italy Next for the West's Populist Wave?", *The Guardian*, November 27, 2016, at https://www.theguardian.com/world/2016/nov/27/matteo-renzi-politics-italy-european-union-brexit-trump.
14. *Outer Space Treaty,* Art. XVI.
15. Richard Sambrook , "Why Big Media Got Trump and Brexit So Wrong", *The Intellectual Takeout,* November 14, 2016, athttp://www.intellectualtakeout.org/blog/why-big-media-got-trump-and-brexit-so wrong; Pankaj Mishra, "How Rousseau Predicted Trump", *The New Yorker*, August 1, 2016, at http://www.newyorker.com/magazine/2016/08/01/how-rousseau-predicted-trump. Mishra reminds that Jean-Jacques Rousseau "warned that the bourgeois values of wealth, vanity, and ostentation would impede rather than advance the growth of equality, morality, dignity, freedom, and compassion. He believed that a society based on envy and the power of money, though it might promise progress, would actually impose psychologically debilitating change on its citizens".
16. "It Is Still 3 Minutes to Midnight", *Bulletin of Atomic Scientists*, at http://thebulletin.org/timeline.
17. Contrary to *Outer Space Treaty,* Art. IV (para.1).
18. Contrary to *Outer Space Treaty,* Art. II.
19. Contrary to *Outer Space Treaty,* Arts. IV and IX.
20. Contrary to *Outer Space Treaty,* Art. II.
21. Contrary to *Outer Space Treaty,* Art. VII.

22. For detailed discussion, see Ram S. Jakhu and Steven Freeland, "The Relationship between the Outer Space Treaty and Customary International Law", accepted for publication in *Proceedings of the International Institute of Space Law*, 2016; Ram S. Jakhu, "Sixty Years of Development of International Space Law", in Stephan Hobe (ed.) *Air Law, Space Law, Cyber Law – the Institute of Air and Space Law at Age 90,* Carl Heymanns Verlag, Cologne, 2016, pp. 75-108; Ram Jakhu, "Legal Issues Relating to the Global Public Interest in Outer Space", *Journal of Space Law*, 32, 2006, pp. 31-110.
23. *Treaty between The United States of America and The Union of Soviet Socialist Republics on The Limitation of Anti-Ballistic Missile Systems, May 26, 1972, 944 UNTS 13 (entered into force October 3, 1972)* [*ABM Treaty*]. On December 13, 2001, the US indicated its intent to withdraw from the Treaty, and its withdrawal became effective six months later, thereby the treaty ceased to be in force.
24. *Strategic Stability Cooperation Initiative*, September 6, 2000, at fas.org/nuke/control/detarget/news/70900.htm.
25. See e.g. Russian Federation, *Reviewing Opportunities for Achieving the Vienna Consensus on Space Security Encompassing Several Regulatory Domains*, UN Doc A/AC.105/C.1/2016/CRP.15 (2016), para. 7:

    The United States vividly demonstrated a connection between diminishing the Committee's role and powers, on the one hand, and manifestations of total disrespect for international law order, on the other, by adopting the commercial space launch competitiveness act on 25 November 2015.

Conclusion

The concept of space governance has been around approximately since the launch of the first artificial satellite in 1957. Efforts have been made by the United Nations (UN) to evolve a nuanced globally accepted mechanism for the conduct of activities in space since the late 1950s. The concept of governance in space has different shades of understanding for various states and even individuals. This book argues that governance mechanisms essentially aims to promote the security, prosperity and values of states through multilateral cooperation and collective actions to safeguard and optimise the use of space as a global commons. It is well understood that collective action and shared strategic goals among states are inextricably tied to international space governance; treaties and agreements, norms, intergovernmental organisations involved in rule-making, monitoring capabilities and joint decision-making advance governance.

During the last 50 years there have been continuous efforts at various levels to ensure that activities in space are conducted in a peaceful manner. Some of these efforts have acquired global acceptability while some continue to remain at the level of continuing debate. On December 22, 2007, the United Nations General Assembly (UNGA) has endorsed the Space Debris Mitigation Guidelines. The UNGA had also adopted a Russian resolution, 'No first Placement of Weapons in Outer Space', during a plenary meeting on December 7, 2015, while the governments of Russia and China are not able to press on their draft Treaty on the Prevention of the Placement of Weapons in Outer Space, the Threat or Use of Force against Outer Space Objects (PPWT). Similarly, the efforts by the European Union (EU) towards the development of the International Code of Conduct for Outer Space Activities (ICoC) still remain inconclusive in spite of wideranging discussions.

The ethos of the Outer Space Treaty (OST) is about maintaining freedom of exploration of space and use of outer space for peaceful purposes and ensuring that no harmful contamination of outer space takes place. It is about ensuring sovereignty and providing adequate safeguards against damage and a redressal mechanism for liabilities. In this context, it is argued that over the last five decades, largely, the OST has stood the test of time. Presently, more than 100 states in the world have pledged to abide by it. Importantly, this mechanism stands accepted by the states having major stakes in the arena of outer space. Legally these states are bound by the provisions of this treaty mechanism, and most of them have followed this treaty mechanism in true spirit. At the same time, during the last five decades there have been some instances which could be categorised as events disturbing the 'peace in space'. Technically, such acts cannot be identified as treaty violations, because the treaty has vagueness in provisions on weaponisation of outer space. Nevertheless, it is mostly argued that this treaty has not been dishonoured till date. That being said, one would have to admit that the treaty still has certain loopholes and limitations that stem from its inability to address 21st century challenges. Hence, it is important to know: 'Is this mechanism adequate enough to handle contemporary and futuristic challenges?' Moreover, there is a need to assess the journey of this treaty during last 50 years and how this mechanism has been practically enforced for the last five decades. The various chapters in this book have essentially attempted to respond to such queries. The book also attempts to identify these modern-day challenges and propose options. The following paragraphs identify some such challenges and limitations and the associated prescriptions and solutions.

Challenges and Limitations

- The OST has provided the requisite parameters for protecting outer space as a global commons, and while those parameters are still appropriate some 50 years later, there is a need for supplementary rules, norms and legal provisions.
- Five decades back, it was not possible to visualise certain contingencies. Major deficiencies that can be observed in the treaty today constitute a mix of unforeseen developments, both at the geopolitical and technological levels. In the majority of cases, space programme initiatives came about only after the adoption of the OST. Now, owing to leapfrogging of techno-political challenges, it has become critical to find a way out by drawing consensus among negotiating members to expand the scope of the OST.

- Challenges owing to overcrowded orbits, space debris, growing demand for scarce radio spectrum particularly for military and cybersecurity, increased space traffic and widened use of space technologies for military purposes demand long-term solutions. It is important to recognise the utilisation of space for military purposes, which has been already ushered in through high resolution imaging, secured communications, PSTN services; however, there is a need to ensure that weaponisation of space does not happen.
- The inherent limitation of the OST is that it is a treaty about not permitting the use of weapons of mass destruction (WMD) in outer space. Therefore, conceptually, from a legal standpoint, outer space remains unprotected from attempts at turning it into a new arena of military confrontation.
- Globally, there are few international bodies that discuss, negotiate and implement different aspects of space governance at the policy levels. However, they are mainly rooted in a model of self-restraint and self-regulation. By itself, the OST regime lacks formal institutional mechanisms to promote international governance for the peaceful uses of space, monitor compliance and make collective decisions about the application of rules.
- Emerging states actors in space view the OST as a legal shield to prevent any dominance by major stakeholder states. But, with increased private sector involvement, compliance with the OST may not be that easy, particularly in the absence of a well-defined legislative mechanism. Also, there exists a wide gap between the requirements of the OST and the existing legal regime for space activities in general.
- The mitigation of space debris is a key area of concern for every state, and the provisions made by the treaty to combat this issue are limited and incapable of a holistic vision.

Recommendations

Fifty years is a sufficient period of time to develop a well-rounded perspective regarding the usefulness of a treaty mechanism and appreciate its merits and demerits. Therefore, assimilating the OST's limitations with changing times and factoring in the possible growth in space technologies, the book puts forward several suggestions to make the treaty mechanism more relevant in today's times. Some plausible options are enumerated as follows:

- It is important to ensure that the efficacy and continuation of the OST, even in its current form, must not be taken for granted. For the present and the future this foundational piece of international legislation is indispensable for the possible future benefits that might be reaped from space and shared by humanity equitably on earth.

- For the last five decades the treaty has proved its merit and has been incredibly useful. It has served its purpose well, and simply by the logic of extrapolation, for the time being, maintaining the status quo would be the best option.
- On the other hand, there are also arguments for a need to contemporise the treaty by including special protocols to the treaty mechanism.
- It is important for the UN to redouble efforts to arrive at a consensus regarding the definitions of key terms. This will help lend clarity and legal certainty, and ensure accessibility and affordability to reliable information.
- States are required to devise their own national legislations in order to supplement the existing treaty mechanisms. Such legislations should address issues of authorisation, continuing supervision, registration of space objects, liability for damages, environmental safeguards and other legal issues flowing from the OST.
- Article IV is one of the important articles of this treaty. It could be considered as both a strength and a weakness of the treaty. The article caters to various key arms control provisions; however, it prohibits only WMDs in space, and not weapons in general. Hence, issues related to conventional kinetic, directed energy weapons (DEW) and the so-called below-WMD level bombs remain unattended. In the absence of definition for 'space Weapons', any small/micro satellite could cause damage to operating satellites, if it is desired so. The OST provides parameters for space activities appropriate for the time and technology of the 1960s. Hence, there is a need to strengthen Article IV in particular, while also formulating supplementary rules, norms and laws to catch up to the technological progress that has taken place since.
- Articles dealing with issues like non-appropriation of celestial bodies could be interpreted differently by different agencies. For example, Article II is interpreted that only State parties only are prohibited from appropriation and not the private parties under their jurisdiction. Issues like space mining, space tourism and the colonisation of the Moon/Mars have highlight OST's limitations. Hence, Article II needs to be strengthened further.

- States need to guard against space "exploration" transforming into space "exploitation". There needs to be a more detailed framework addressing issues that emerge from private enterprises' increased enthusiasm for foraying into space.

The Way Ahead

It is important to undertake the examination of the policy effectiveness of the OST in the face of modern challenges. For the last few years, serious and sincere efforts are being made to establish a legal regime for activities in outer space. The debate between soft law and hard law has also continued for close to a decade. However, none of these efforts are moving towards establishing global acceptability for the formulation of any globally accepted mechanism. The need of the hour is to have a legally binding mechanism; however, efforts towards establishing even a voluntary and not legally binding mechanism have not succeeded thus far. Hence, under such circumstances major spacefaring nations need to figure out minimally acceptable standards which could lead to defining certain norms of responsible behaviour.

Reviewing and fine-tuning the OST is the ideal situation, but the current political scenario does not appear ready for any meaningful engagement. Therefore, it is important for states to consider a few practical means to dispel some of the prevailing mistrust – stemming from rivalry and competition – in order to develop predictable and sustainable options.

- To ensure strategic stability, there is a need for optimal collective action through the establishment of acceptable governance mechanisms.
- A mechanism including transparency and confidence building measures (TCBMs) could be put in place. Such instruments are essentially voluntary, non-legal measures that promote mutual trust and encourage cooperation and transparency, while reducing wariness, competition and misperceptions as a means to prevent intended or unintended conflicts. The TCBMs could serve as a bridge between the idea of an effective instrument and the fructification of one. The group of governmental experts (GGE) through the UNGA has recommended for a TCBMs mechanism (2013) however, much needs to be done in the regards from the point of view of implementation.
- There is an urgent need to institute multiple levels of dialogue and interaction for a broader understanding of the different orientations and objectives of space programmes and policies. This could be achieved by strengthening existing regional and other multilateral

initiatives addressing issues concerning space security. Furthermore, instituting new initiatives at bilateral, trilateral or quadrilateral levels to work towards establishing arrangements and TCBMs could help mitigate some of the distrust amongst states. The Code of Conduct for outer space activities is in advanced stage, but not seem to be getting consensus.

The Reality

The overall analysis indicates that various articles in the existing treaty mechanism require further strengthening. However, the treaty mechanism does not have any provisions to undertake amendments. Amendments can be made only by consensus under the UN procedures, the method for which generally includes:

(1) To begin with, the relevant issue is raised as an agenda item in the Legal Sub Committee (LSC) of COPOUS.
(2) Then the issue gets debated in the LSC and the amendment is voted upon and adopted.
(3) Subsequently, the LSC COPOUS submits the amended provision of the OST as a Resolution to the UN General Assembly for adoption by consensus.

At the moment, states (both space-faring and others) do not seem to have a desire towards developing or amending such a mechanism concerning long-term sustainability of outer space activities and space security. Hence, in the near future, any tweaking of the OST appears unlikely. The future of the OST depends on the sincerity of the signatory states and they must work to protect it continuously. Presently, it is important to ensure that OST gets strengthened further by developing various additional soft-law mechanisms. Even 50 years' ago, this treaty had catered for various challenges which are emerging at present like activities of non-governmental entities in outer space, including the moon and other celestial bodies (refer Article VI). In short, the Outer Space Treaty has stood the test of time.

APPENDICES

Appendix I

The Text of Outer Space Treaty

TREATY ON PRINCIPLES GOVERNING THE ACTIVITIES OF STATES IN THE EXPLORATION AND USE OF OUTER SPACE, INCLUDING THE MOON AND OTHER CELESTIAL BODIES

The States Parties to this Treaty,

Inspired by the great prospects opening up before mankind as a result of man's entry into outer space,

Recognizing the common interest of all mankind in the progress of the exploration and use of outer space for peaceful purposes,

Believing that the exploration and use of outer space should be carried on for the benefit of all peoples irrespective of the degree of their economic or scientific development,

Desiring to contribute to broad international co-operation in the scientific as well as the legal aspects of the exploration and use of outer space for peaceful purposes,

Believing that such co-operation will contribute to the development of mutual understanding and to the strengthening of friendly relations between States and peoples,

Recalling resolution 1962 (XVIII), entitled "Declaration of Legal Principles Governing the Activities of States in the Exploration and Use of Outer Space", which was adopted unanimously by the United Nations General Assembly on 13 December 1963,

Recalling resolution 1884 (XVIII), calling upon States to refrain from placing in orbit around the earth any objects carrying nuclear weapons or any other kinds of weapons of mass destruction or from installing such weapons on celestial bodies, which was adopted unanimously by the United Nations General Assembly on 17 October 1963,

Taking account of United Nations General Assembly resolution 110 (II) of 3 November 1947, which condemned propaganda designed or likely to provoke or encourage any threat to the peace, breach of the peace or act of aggression, and considering that the aforementioned resolution is applicable to outer space,

Convinced that a Treaty on Principles Governing the Activities of States in the Exploration and Use of Outer Space, including the Moon and Other Celestial Bodies, will further the purposes and principles of the Charter of the United Nations,

Have agreed on the following:

Article I

The exploration and use of outer space, including the moon and other celestial bodies, shall be carried out for the benefit and in the interests of all countries, irrespective of their degree of economic or scientific development, and shall be the province of all mankind.

Outer space, including the moon and other celestial bodies, shall be free for exploration and use by all States without discrimination of any kind, on a basis of equality and in accordance with international law, and there shall be free access to all areas of celestial bodies.

There shall be freedom of scientific investigation in outer space, including the moon and other celestial bodies, and States shall facilitate and encourage international co-operation in such investigation.

Article II

Outer space, including the moon and other celestial bodies, is not subject to national appropriation by claim of sovereignty, by means of use or occupation, or by any other means.

Article III

States Parties to the Treaty shall carry on activities in the exploration and use of outer space, including the moon and other celestial bodies, in accordance with international law, including the Charter of the United Nations, in the interest of maintaining international peace and security and promoting international co-operation and understanding.

Article IV

States Parties to the Treaty undertake not to place in orbit around the earth any objects carrying nuclear weapons or any other kinds of weapons of mass destruction, install such weapons on celestial bodies, or station such weapons in outer space in any other manner.

The moon and other celestial bodies shall be used by all States Parties to the Treaty exclusively for peaceful purposes. The establishment of military bases, installations and fortifications, the testing of any type of weapons and the conduct of military manoeuvres on celestial bodies shall be forbidden. The use of military personnel for scientific research or for any other peaceful purposes shall not be prohibited. The use of any equipment or facility necessary for peaceful exploration of the moon and other celestial bodies shall also not be prohibited.

Article V

States Parties to the Treaty shall regard astronauts as envoys of mankind in outer space and shall render to them all possible assistance in the event of accident, distress, or emergency landing on the territory of another State Party or on the high seas. When astronauts make such a landing, they shall be safely and promptly returned to the State of registry of their space vehicle.

In carrying on activities in outer space and on celestial bodies, the astronauts of one State Party shall render all possible assistance to the astronauts of other States Parties.

States Parties to the Treaty shall immediately inform the other States Parties to the Treaty or the Secretary-General of the United Nations of any phenomena they discover in outer space, including the moon and other celestial bodies, which could constitute a danger to the life or health of astronauts.

Article VI

States Parties to the Treaty shall bear international responsibility for national activities in outer space, including the moon and other celestial bodies, whether such activities are carried on by governmental agencies or by non-governmental entities, and for assuring that national activities are carried out in conformity with the provisions set forth in the present Treaty. The activities of non-governmental entities in outer space, including the moon and other celestial bodies, shall require authorization and continuing supervision by the appropriate State Party to the Treaty. When activities are carried on in outer space, including the moon and other celestial bodies, by an international organization, responsibility for compliance with this Treaty shall be borne both by the international organization and by the States Parties to the Treaty participating in such organization.

Article VII

Each State Party to the Treaty that launches or procures the launching of an object into outer space, including the moon and other celestial bodies, and each State Party from whose territory or facility an object is launched, is internationally liable for damage to another State Party to the Treaty or to its natural or juridical persons by such object or its component parts on the Earth, in air or in outer space, including the moon and other celestial bodies.

Article VIII

A State Party to the Treaty on whose registry an object launched into outer space is carried shall retain jurisdiction and control over such object, and over any personnel thereof, while in outer space or on a celestial body. Ownership of objects launched into outer space, including objects landed or constructed on a celestial body, and of their component parts, is not affected by their presence in outer space or on a celestial body or by their return to the Earth. Such objects or component parts found beyond the limits of the State Party to the Treaty on whose registry they are carried shall be returned to that State Party, which shall, upon request, furnish identifying data prior to their return.

Article IX

In the exploration and use of outer space, including the moon and other celestial bodies, States Parties to the Treaty shall be guided by the principle of co-operation and mutual assistance and shall conduct all their activities in outer space, including the moon and other celestial bodies, with due regard to the corresponding interests of all other States Parties to the Treaty. States Parties to the Treaty shall pursue studies of outer space, including the moon and other celestial bodies, and conduct exploration of them so as to avoid their harmful contamination and also adverse changes in the environment of the Earth resulting from the introduction of extraterrestrial matter and, where necessary, shall adopt appropriate measures for this purpose. If a State Party to the Treaty has reason to believe that an activity or experiment planned by it or its nationals in outer space, including the moon and other celestial bodies, would cause potentially harmful interference with activities of other States Parties in the peaceful exploration and use of outer space, including the moon and other celestial bodies, it shall undertake appropriate international consultations before proceeding with any such activity or experiment. A State Party to the Treaty which has reason to believe that an activity or experiment planned by another State Party in outer space, including the moon and other celestial bodies, would cause potentially harmful interference with activities in the peaceful exploration and use of outer space, including the moon and other celestial bodies, may request consultation concerning the activity or experiment.

Article X

In order to promote international co-operation in the exploration and use of outer space, including the moon and other celestial bodies, in conformity with the purposes of this Treaty, the States Parties to the Treaty shall consider on a basis of equality any requests by other States Parties to the Treaty to be afforded an opportunity to observe the flight of space objects launched by those States. The nature of such an opportunity for observation and the conditions under which it could be afforded shall be determined by agreement between the States concerned.

Article XI

In order to promote international co-operation in the peaceful exploration and use of outer space, States Parties to the Treaty conducting activities in outer space, including the moon and other celestial bodies, agree to inform the Secretary-General of the United Nations as well as the public and the international scientific community, to the greatest extent feasible and practicable, of the nature, conduct, locations and results of such activities. On receiving the said information, the Secretary-General of the United Nations should be prepared to disseminate it immediately and effectively.

Article XII

All stations, installations, equipment and space vehicles on the moon and other celestial bodies shall be open to representatives of other States Parties to the Treaty on a basis of reciprocity. Such representatives shall give reasonable advance notice of a projected visit, in order that appropriate consultations may be held and that maximum precautions may betaken to assure safety and to avoid interference with normal operations in the facility to be visited.

Article XIII

The provisions of this Treaty shall apply to the activities of States Parties to the Treaty in the exploration and use of outer space, including the moon and other celestial bodies, whether such activities are carried on by a single State Party to the Treaty or jointly with other States, including cases where they are carried on within the framework of international intergovernmental organizations.

Any practical questions arising in connection with activities carried on by international intergovernmental organizations in the exploration and use of outer space, including the moon and other celestial bodies, shall be resolved by the States Parties to the Treaty either with the appropriate international organization or with one or more States members of that international organization, which are Parties to this Treaty.

Article XIV

1. This Treaty shall be open to all States for signature. Any State which does not sign this Treaty before its entry into force in accordance with paragraph 3 of this article may accede to it at anytime.
2. This Treaty shall be subject to ratification by signatory States. Instruments of ratification and instruments of accession shall be deposited with the Governments of the United Kingdom of Great Britain and Northern Ireland, the Union of Soviet Socialist Republics and the United States of America, which are hereby designated the Depositary Governments.
3. This Treaty shall enter into force upon the deposit of instruments of ratification by five Governments including the Governments designated as Depositary Governments under this Treaty.
4. For States whose instruments of ratification or accession are deposited subsequent to the entry into force of this Treaty, it shall enter into force on the date of the deposit of their instruments of ratification or accession.
5. The Depositary Governments shall promptly inform all signatory and acceding States of the date of each signature, the date of deposit of each instrument of ratification of and accession to this Treaty, the date of its entry into force and other notices.

6. This Treaty shall be registered by the Depositary Governments pursuant to Article 102 of the Charter of the United Nations.

Article XV

Any State Party to the Treaty may propose amendments to this Treaty. Amendments shall enter into force for each State Party to the Treaty accepting the amendments upon their acceptance by a majority of the States Parties to the Treaty and thereafter for each remaining State Party to the Treaty on the date of acceptance by it.

Article XVI

Any State Party to the Treaty may give notice of its withdrawal from the Treaty one year after its entry into force by written notification to the Depositary Governments. Such withdrawal shall take effect one year from the date of receipt of this notification.

Article XVII

This Treaty, of which the English, Russian, French, Spanish and Chinese texts are equally authentic, shall be deposited in the archives of the Depositary Governments. Duly certified copies of this Treaty shall be transmitted by the Depositary Governments to the Governments of the signatory and acceding States.

IN WITNESS WHEREOF the undersigned, duly authorized, have signed this Treaty.

DONE in triplicate, at the cities of London, Moscow and Washington, the twenty-seventh day of January, one thousand nine hundred and sixty-seven.

Note

1. *Official Records of the General Assembly, Twenty-first Session*, agenda items 30, 89 and 91, document A/6431.

Reference: http://www.unoosa.org/oosa/en/ourwork/spacelaw/treaties/outerspacetreaty.html

APPENDIX II

The Outer Space Treaty at a Glance

The 1967 Outer Space Treaty bans the stationing of weapons of mass destruction (WMD) in outer space, prohibits military activities on celestial bodies, and details legally binding rules governing the peaceful exploration and use of space.

Treaty Terms

The treaty forbids countries from deploying "nuclear weapons or any other kinds of weapons of mass destruction" in outer space. The term "weapons of mass destruction (WMD)" is not defined, but it is commonly understood to include nuclear, chemical, and biological weapons. The treaty, however, does not prohibit the launching of ballistic missiles, which could be armed with WMD warheads, through space. The treaty repeatedly emphasizes that space is to be used for peaceful purposes, leading some analysts to conclude that the treaty could broadly be interpreted as prohibiting all types of weapons systems, not just WMD, in outer space.

The treaty's key arms control provisions are in Article IV. States-parties commit not to:

- Place in orbit around the Earth or other celestial bodies any nuclear weapons or objects carrying WMD.
- Install WMD on celestial bodies or station WMD in outer space in any other manner.
- Establish military bases or installations, test "any type of weapons," or conduct military exercises on the moon and other celestial bodies.

Other treaty provisions underscore that space is no single country's domain and that all countries have a right to explore it. These provisions state that:

- Space should be accessible to all countries and can be freely and scientifically investigated.

- Space and celestial bodies are exempt from national claims of ownership.
- Countries are to avoid contaminating and harming space or celestial bodies.
- Countries exploring space are responsible and liable for any damage their activities may cause.
- Space exploration is to be guided by "principles of cooperation and mutual assistance," such as obliging astronauts to provide aid to one another if needed.

Like other treaties, the Outer Space Treaty allows for amendments or member withdrawal. Article XV permits countries to propose amendments. An amendment can only enter into force if accepted by a majority of states-parties, and it will only be binding on those countries that approve the amendment. Article XVI states a country's withdrawal from the treaty will take effect a year after it has submitted a written notification of its intentions to the depositary states-the United States, Russia, and the United Kingdom.

History

Talks on preserving outer space for peaceful purposes began in the late 1950s at the United Nations. The United States and its Western allies submitted proposals in 1957 on reserving space exclusively for "peaceful and scientific purposes," but the Soviet Union rejected these efforts because it was preparing to launch the world's first satellite and test its first intercontinental ballistic missile.

In 1963, the UN General Assembly approved two resolutions on outer space that subsequently became the basis for the Outer Space Treaty. UN Resolution 1884 called on countries to refrain from stationing WMD in outer space. UN Resolution 1962 set out legal principles on outer space exploration, which stipulated that all countries have the right to freely explore and use space.

The United States and Soviet Union submitted separate draft outer space treaties to the UN General Assembly in June 1966. A mutually agreed treaty text was worked out over the next six months, and the UN General Assembly gave its approval of the treaty on December 19, 1966. The treaty opened for signature in Washington, Moscow, and London on January 27, 1967 and entered into force October 10, 1967.

Reference: Researched by Ben Rusek, A *Fact Sheet* by Arms Control Association, available at https://www.armscontrol.org/factsheets/outerspace

Index